# YouTube Marketing

How to go viral, growing followers, become an influencer and make money

By Dave Miller

© **Copyright 2019 by Dave Miller - All rights reserved.**

This document is geared towards providing exact and reliable information in regards to the topic and issue covered. The publication is sold with the idea that the publisher is not required to render accounting, officially permitted, or otherwise, qualified services. If advice is necessary, legal or professional, a practiced individual in the profession should be ordered.

- From a Declaration of Principles which was accepted and approved equally by a Committee of the American Bar Association and a Committee of Publishers and Associations.

In no way is it legal to reproduce, duplicate, or transmit any part of this document in either electronic means or in printed format. Recording of this publication is strictly prohibited and any storage of this document is not allowed unless with written permission from the publisher. All rights reserved.

The information provided herein is stated to be truthful and consistent, in that any liability, in terms of inattention or otherwise, by any usage or abuse of any policies, processes, or directions contained within is the solitary and utter responsibility of the recipient reader. Under no circumstances will any legal responsibility or blame be held against the publisher for any reparation, damages, or monetary loss due to the information herein, either directly or indirectly.

Respective authors own all copyrights not held by the publisher.

The information herein is offered for informational purposes solely, and is universal as so. The presentation of the information is without contract or any type of guarantee assurance.

The trademarks that are used are without any consent, and the publication of the trademark is without permission or backing by the trademark owner. All trademarks and brands within this book are for clarifying purposes only and are the owned by the owners themselves, not affiliated with this document.

## Table of contents

### CHAPTER 1: GETTING TO KNOW YOUTUBE MARKETING ............................................................................. 9

1.1 Introduction ................................................................. 9

1.2 Three main pillars .................................................... 11

1.3 Popular Companies promoting business through YouTube ......... 12

### CHAPTER 2: YOUTUBE CHANNEL ......................................... 16

2.1 What is a YouTube channel? .................................. 16

2.2 What is YouTube influencer ................................... 16

2.3 Creating a YouTube Channel ................................. 17

2.4 How YouTube Works .............................................. 20

2.5 Checking your Audience ........................................ 22

2.6 Promotion of YouTube Channel ............................. 28

### CHAPTER 3: MONETIZING YOUR YOUTUBE CHANNEL . 32

3.1 Four ways to monetize your channel .................... 35

### CHAPTER 4: TYPES OF YOUTUBE ADVERTISEMENTS .. 40

4.1 True View Advertisement ....................................... 40

4.2 Non-Skippable YouTube ads ................................................................ 43

4.3 Bumper Ads ........................................................................................ 44

4.4 Display ads ........................................................................................ 45

4.6 Sponsored cards ............................................................................... 46

4.7 Masthead format ............................................................................... 46

## CHAPTER 5: MAINTAINING YOUR CHANNEL ............... 49

5.1 Posting material multiple times a week ........................................... 49

5.2 Develop Sustainable Video Work Flow ........................................... 49

5.3 Adding Interesting Hook .................................................................. 50

5.4 Keeping Title and Openings Short ................................................... 51

5.5 Adding End Screens ......................................................................... 51

5.6 Edit Distractions from Videos .......................................................... 52

5.7 Thumbnails to click on ..................................................................... 53

5.8 Tactics from top-performing Videos ................................................ 56

5.9 Making Long Videos to improve watch time .................................. 56

5.10 Going Live on YouTube ................................................................. 57

5.11 Using Videos in Content Mix ......................................................... 58

5.12 Make Videos in Series .................................................................... 59

5.13 Collaboration with other You Tubers ................................................. 60

## CHAPTER 6: SEO (SEARCH ENGINE OPTIMIZATION) ON YOUTUBE VIDEOS ........................................................................... 64

6.1 YouTube Keyword Search ................................................................. 64

6.2 Inserting your Keyword Naturally ...................................................... 65

6.3 Optimize the definition of your video. ............................................... 66

6.4 Tagging Video with Proper Keywords ............................................... 67

6.5 Categorizing Your Video Content ..................................................... 68

6.6 Customizing your Thumbnail for Video ............................................ 68

6.7 Using SRT Files to Add subtitles & closed captions ....................... 69

6.8 Adding Cards & End Screens to increase your YouTube Viewership ................................................................................................ 70

6.9 YouTube SEO Tools .......................................................................... 72

## CHAPTER 7: BEST YOUTUBE CHANNEL IDEAS ............... 76

## CHAPTER 8: YOUTUBE MARKETING IN MAIN SECTORS OF LIFE ................................................................................................... 110

8.1 Education ........................................................................................... 110

8.2 HealthCare ......................................................................................... 117

8.3 Agriculture ..................................................................................128

## CHAPTER 9: ROLE OF SOCIAL MEDIA ON YOUTUBE MARKETING ................................................................................ 139

9.1 Social Media Marketing is good or bad ............................................142

9.2 How to Use YouTube and Instagram to Establish Authority ........144

## CHAPTER 10: FUTURE OF YOUTUBE MARKETING ....... 148

10.1 What's next? ..................................................................................148

10.2 Video content marketing revolution ...............................................152

## 11: CONCLUSION .................................................................. 158

## 12. REFERENCES .................................................................. 162

# Chapter 1: Getting to Know YouTube Marketing

## 1.1 Introduction

New forms of marketing are growing across channels, and the industry is diversifying. There is good news as well as bad news for marketers. Marketing diversification will be either a new opportunity or an external obstacle for marketers, depending on how marketers react and adapt to the industry's changes. Using multiple media channels to spread your content and grow your brand will become more important than ever. It used to be fairly easy to build a web page. But now, because of the enormous amount of content and information available at your disposal, the Internet has become an incredibly competitive area.

No one could deny that over a couple of years, video marketing has increased its influence on an online marketing platform. While social platforms such as Facebook, Instagram, Twitter, Pinterest, and Snap Chat have become great social media marketing platforms, but YouTube still shares an enormous vacuum with almost 2 billion users worldwide. It is the second in popularity for Facebook, with more than 2.32 billion users worldwide. According to an estimate, YouTube viewers are spending 180 million hours watching video content every day.

YouTube Marketing is a crucial strategy for both internet marketers and online business owners to benefit from the web's massive shift to video. Only the fact that YouTube Advertising is a growing trend and a very successful way to reach your target market is explained by the huge traffic received by this platform each day.

Not only is your audience on YouTube, but YouTube will help improve your SEO and overall brand visibility as the second largest search engine on the internet. YouTube provides advertisers with unique content that is easy to consume and distribute with audiences.

YouTube ads can be a device for advertisers to threaten. To YouTube pros and beginners alike, that's why we built this complete guide. Below we will walk through each marketing phase on YouTube — from building a YouTube channel and optimizing SEO videos to running YouTube campaign and analysing video analytics.

In sales, certain aspects are not crystal clear when it comes to calculating performance. While some campaigns provide a clear roadmap for their success or failure, others may become quite complicated. Marketing material is an environment that tends to lean toward the latter. It does work, though, but takes a majority of the time.

It is difficult to quantify content marketing because it does not always explicitly translate into a sale. If someone reads and comments on an article you wrote and shares it, they have not brought the full circle of the sales cycle. Their acts have just broadened your content's scope. While this is a critical component of any marketing strategy, to measure outcomes, it does not provide her data.

Think of marketing content as a means to an end. Ultimately, the fruits of your labour will pay off. It only takes time and a good amount of know-how. Marketing's three foundations are engaging, educating, and empowering people. If you do these things correctly, your company will expand. Here is a further overview of how these three key elements specifically affect the effectiveness of content marketing.

## 1.2 Three main pillars

### Engage

You must first grasp them in order to engage with your target audience. What kind of material are they reacting to? Which products are you providing in accordance with your interests? What is the use of social media sites? Buyers' responses may differ depending on a variety of factors such as age, sex, geographic location, level of education, and employment. It is important to provide content in a way that resonates with the target audience.

### Educate

You must first grasp them in order to engage with your target audience. What kind of material are they reacting to? Which products are you providing in accordance with your interests? What is the use of social media sites? Buyers' responses may differ depending on a variety of factors such as age, sex, geographic location, level of education, and employment. It is important to provide content in a way that resonates with the target audience.

**Empower**

Marketing content empowers consumers to make decisions that are more informed. The internet is full of data, and not all of it should be trusted, unfortunately. People often don't know their options and make less than ideal decisions because of this. This can be devastating for health care. You empower consumers with the knowledge and confidence they need to make the best purchasing decisions for their situation when you succeed in content marketing.

## 1.3 Popular Companies promoting business through YouTube

Studies show that viewing video is one of the Smartphone users' most common activities. Just under half of them watch videos on their phones regularly.

**Play Station**

Adweek named Sony's YouTube PlayStation channel one of the top 10 product channels, so it's no brainer that any organization will learn a lot of insight from its nearly 3,500 videos. One thing PlayStation does consistently is to include its logo in its content's left corner, reinforce the branding of its company, and remind users who created the video.

**Walmart**

Many people have strong reactions to Walmart, but the quality of content on their YouTube channel cannot be claimed to be remarkable. They not only include easy recipes; they also have testimonials on price matching, new product tutorials, and blogger partners who contribute their own product reviews.

**Red bull**

Since Red Bull is an energy drink, their marketing revolves around encouraging a healthy, energy-filled lifestyle. Their YouTube channel has more than 4 million subscribers, and in 6 years, they have exceeded a billion video views. Their most popular videos are full of athletes, daredevils, and other risk-takers who do great things like climbing frozen Niagara Falls, building a wooden bicycle.

**Blendtec**

For their first, would it Blend, Blendtec became extremely popular? Videos back when new iPhones just arrived on the market were mixed. This costly (and destructive) stunt demonstrated their blenders' strength and they now have nearly 775,000 channel subscribers, and many of their videos have almost a million views.

## The Ellen Show

While the daily talk show of Ellen DeGeneres is not a company. Her YouTube channel does a fantastic job of segmenting what she does (e.g., the bits she airs) into separate YouTube videos every day, while also uniting video marketing with other social media channels. They posted a new video on Twitter in the following example, tagging the other users in the video and providing a brief overview of what it was.

## Ramit Sethi

Ramit is a writer and researcher in the field of personal finance and entrepreneurship. Although he's not a corporation like The Ellen Show (he's more of a thought leader building a business around what he's talking about), but his YouTube channel has a lot of good ideas that companies may benefit from. He breaks his videos into specific sections, offering the variety that users want (as mentioned above), while at the same time being more organized to avoid overwhelming users. You will see categories ranging from "Tell Ramit" to wage negotiation tactics on his main channel page.

# Chapter 2: YouTube Channel

## 2.1 What is a YouTube channel?

Anyone joining YouTube as a member will access a private YouTube channel. The stream serves as the user account home page. The channel shows the account name, personal profile, public videos uploaded by the member, and any user information entered by the member after the user enters and approves the information. If you are a member of YouTube, you can customize your personal channel's context and colour scheme and monitor some of the details shown on it. Businesses can have outlets, as well. Such networks vary from private channels because they are capable of having more than one owner or administrator. A member of YouTube can use a Brand Account to open a new business site.

## 2.2 What is YouTube influencer

YouTube influencer is someone who creates a significant follow-up on YouTube's video platform, helping to set trends and provide data, for example, to others who are looking to buy a specific product or service. Some of the most popular influencers on YouTube are a mixture of performing artists, visual artists (including makeup artists), or some athletes. They can share their skills and make them easily accessible by making videos on YouTube.

## 2.3 Creating a YouTube Channel

Setting up your own YouTube channel takes just a few minutes. Then you want to customize the channel by changing a few settings, uploading your image or logo, and connecting your channel to other online social networking sites, such as Facebook, Twitter, and Google+.

First of all, you need to build your free account for Google. If you are starting a YouTube channel for your company, use a specific and non-personal email address to set up a separate Google / YouTube account from scratch. This way, someone else inside the company will manage the network without giving out the username and password to Google's personal account. Note that only one YouTube channel can be connected to each Google account.

There is no such thing as a dedicated business account or a corporate YouTube channel. You will need to customize the settings of a standard YouTube channel to best suit your audience and show your company, image and brand, and videos.

### Creating a Gmail Account

Follow these steps to create a unique Google account:

**1.** Launch any internet-connected web browser on your computer and visit www.youtube.com.

**2.** Tap on the "Sign In" button from the YouTube website, which is displayed near the top-right corner of the screen.

**3.** When the "Sign In" window appears on YouTube (with the Google logo in the top-left corner), press the "Create an Account" button in the top-right corner. You will be asked to create a new Google account for the first time.

**4.** Fill in the fields on the screen "Build a New Google Account." You will be asked to enter your name and your first name. You will then be prompted to pick a unique username for Google. First, build and validate your account password, enter your birthday and age, your mobile phone number, and your current email address. For example, if you create a YouTube channel for your company or service when asked for your current email address, do not use a personal email address. From the pull-down menu, pick your location and then agree to the "Terms of Service" on the screen. To continue, press the "Next Move" button.

The username you pick from Google will also become your channel name for YouTube, and the account will receive a free Gmail address. Use your name as your username or make an intelligent choice to communicate with your intended audience. It should be easy to spell the channel name/username, and something people will remember. For example, if your YouTube channel is advertising a product, consider using the product name (assuming that someone other than you or your company is not copyright or marked) as your username.

**5.** First, you need to build your profile for Google Account. It involves uploading an optional photograph of the user. To do this, click the "Add Photo Profile" button, then click the "Next Move" button to proceed. When you create a business or association page, upload a company logo or brand image as opposed to a personal photo or headshot.

**6.** Tap on the "Return to YouTube" button once you have set up your Google account.

**7.** You will receive two emails from Google within a few minutes. You're going to be asked to search your current email address. To do this, click the link given in the message. You will get the second email with information on your existing Gmail account. Save the details for later reference in this message.

**Turn Your Google Account into a YouTube Channel**

You can create your own YouTube channel and easily customize it and then use your Google account to fill it with your videos (which also serves as your video sharing YouTube account). The following steps are to create a free YouTube channel when Google has set up a legitimate account.

1. Visit www.youtube.com and sign in with your Google account's username and password. YouTube's main home screen is shown.

2. You will see your account profile picture near the top-right corner of the screen. Tap it to open the Google Account Menu in the top-right corner of the screen.

3. Click in the top-right portion of the monitor on the "My Network" button. The "Launch Your YouTube Channel" screen will be shown. From this initial screen, you will see your profile photo, username, or first and last name. To change your Google profile, which will be your identity with public online information about you, click the "Edit" link associated with the "From Your Google Profile" option.

4. Under the heading "Activities you're going to share on your channel," you'll see four options, labelled "Like a video," "Comment on a video," "Favorited a video," and "Subscribe to a channel." Fill in the checkbox for each event you want people to be able to do on your YouTube channel page.

5. Click the "Yes, I'm ready to go on" button. The YouTube channel has now been set up. The next step is to start uploading your website's images.

## 2.4 How YouTube Works

YouTube assigns you a private channel when you become a member of YouTube. The channel has sections designed to show a brief personal profile, thumbnails of videos that you have posted, members that you have subscribed to, videos from other members that you have selected as favourites, lists of members that are your friends and subscribers, and a section where other people can comment on your channel.

You can reach another member's channel by clicking on their user name. Here you can see all the YouTuber videos and all the favourites he or she has picked. You can even see the other participants to which YouTuber subscribes. Personal channels allow you to view YouTube as a social network rather than as a simple video server — you'll find people who like the same kinds of videos you watch.

Your channel is a digital wasteland when you first create an account with all pieces that are blank. Luckily, YouTube makes it easy to turn your channel into an appealing online destination. After filling out your profile information, you can change the colour scheme of your site. You can either choose one of YouTube's approved colour schemes or use hexadecimal colour values to create your own.

You can use a simple menu to modify your channel model. You can choose to cover or display pieces, and you can want to reveal them on the left or right side of the web page. These types of options allow a unique flow to be developed.

Once you set up your channel, it's time to fill in those empty fields. Visit the website and select the videos you love. To add the video to your personal channel's "favourites" section, you can watch videos and click the preferred button. You should subscribe to the person who uploaded the clip to keep up with his or her uploads— a screenshot of the member's latest video will appear in your personal stream's "subscriptions" section. If you upload a video of your own (unless you have changed the design options) it will appear in the top right of your page. You will fill in your personal channel's "uploads" tab as you add more videos, and the new clip appears at the top right of your profile.

## 2.5 Checking your Audience

This month, YouTube has reached 1 billion active users–that's almost 1/3 of all people on the internet, spending hundreds of millions of hours on YouTube and creating billions of views!

But in fact, many companies are not investing in YouTube advertising, as video production is more complicated and costlier than writing a blog post. Every other day you can't just crank out loads of awesome videos for your website.

A barrier to entry, however, also means an opportunity gate. Yet YouTube is far from being saturated. You could start on YouTube today and do more for yourself than just great. Nonetheless, it is not as easy as it sounds because if your videos don't hit the right engagement thresholds from the outset, your videos won't go anywhere as the YouTube algorithm will kill your views of the future.

**A LOST VIDEO IS A COSTLY MISTAKE.**

Therefore, before you move into video production, it is wise to spend time identifying, knowing, and reaching your target audience. The essence of how you search your YouTube channel for the right audience. You describe it, understand it, and find it let me guide you through every step.

**WHAT IS THE TARGET AUDIENCE?**

You should think about who your target audience would be before you start working on your next YouTube video. For example, for people who want to lose weight, you want to start a YouTube channel. But this will be a much too broad audience to be considered a "target audience" for your video. Depending on your race, age, occupation, position, level of knowledge, or location, you need to limit your audience.

YouTube removed its keyword tool a while back, so you can use Ad Words & Keyword Planner to get a sense of the monthly search volume for various keywords. However, this is not reliable as it indicates the number of searches on Google search, not a YouTube search.

The more precise you get, the more you can create "targeted" content, which is less efficient.

Now, there's something so familiar about "weight loss." It's got about 5,640,000 YouTube results. To be one of the top YouTube videos for this keyword, you will contend with around

5.6 million clips. Without a massive marketing campaign, you will not be able to rank for this.

Nevertheless, you can easily find and establish a target audience for your next video by looking at suggestions for the YouTube keyword and brainstorming with your ideas.

So, you can describe your target audience for your YouTube channel as women over 40 who want to lose weight, teenage girls who want to lose weight or women with thyroid who want to lose weight. It is termed as 'Target Audience Classes.'

Now, understanding these classes is the second step.

The growing of these groups would have specific priorities, goals, conditions, age, and issues that they are seeking to solve.

Women over 40 are probably those who aren't looking to actively spend a lot of money on weight loss, and are just looking for straightforward information and tips to help them stay in shape.

While the "teenage girls" segment would be the one that will be willing to spend a decent amount of money to get in shape and look good. Through promoting products and solutions to them and moving them from the YouTube page to your website, you can monetize this audience even beyond YouTube ads.

And the women suffering from thyroid would be looking for natural remedies to improve their metabolism. Furthermore, this group will be highly monetizable as they already spend money on their care and wouldn't mind scratching out a few extra bucks for something that improves the method.

Learning to define your different audience types within your market is essential to the success of your YouTube channel. You can easily create video content that helps them achieve what they want and become your loyal subscribers, understanding their unique needs.

Here's how you can decide what content to build for your target audience – "Create your video content at the confluence of what your audience cares about and what your channel stands for." [Tweet this] –What your audience cares about is their passions, interests, problems, and needs.

What your channel is about is the purpose of your network.

Now, the third and final step is to find out where the target audience is staying.

Finding Your Target Audience, Your YouTube videos are mainly accessed from: Inside YouTube (YouTube Traffic) Search Engines (Search Traffic) Other blogs, groups, boards, forums, fan pages and social media sites where your videos can be embedded or references to your videos can be posted (Referral & Social Traffic).

Inside YouTube Participating and communicating inside YouTube is the golden rule.

Commit to post on other YouTube channels and take part in conversations on different channels. You will gain your target audience by commenting on other YouTube channels that they like as well. Going through the feedback on the websites of your competitors and see if any users have not been happy with the video of the competitor and are still looking for answers, here's your opportunity to help and share links to your photos.

**Comment on video and react to popular videos in your niche.**

Try to appeal to the same audience as yours with other YouTube channels, but do not sell the same thing.

Partner with other channels on YouTube, place them in your channel or their videos get featured and invited.

YouTube Channel Optimization makes sure you take care of the fundamentals of YouTube optimization, such as using specific keywords in the video title, Adding appropriate tags to the clip write video description and describing the challenges you're going to solve through your video Adding action comment at the end of all videos to get users to submit.

Search engines are vehicles for fulfilling demand. After querying their specific issue, users will land on your video and therefore come with a higher purpose. All you need to do is boost your search engine rating YouTube videos.

Through targeting phrases with the following keywords in them, you will increase your chances of appearing in searches: Analysis Tutorial Demo Check Video Walkthrough How to Step through Step You will draw more viewers to your videos if you post videos on current and popular topics utilizing QDF algorithms. When it comes to an existing topic, the keyword is warm and straightforward to rank for. If you get it right, this strategy can sometimes push traffic truckloads to your YouTube video.

Different blogs, groups, boards, forums, fan pages, QA platforms, and social media Post on websites such as Quora, Twitter, fan communities and engage in discussions relevant to the interests of your target audience. In meetings, pre-promote your clips and ask participants if they want to watch a video that you're working on. They're going to watch it and encourage it in their circle if they show interest right now.

When you spend 20% of your time making an impressive clip, you will pay 80% of your time researching your target audience and their pre-promotion before that. (Tweet this) Final Takeaway Demographics is one more thing that adds to the pressure of finding a target audience. Although insights such as age, sex, and location may be useful, they are often superficial and can't tell what your audience is battling with. This is that something you need to protect against.

Reflect on what you want to do differently, what motivates you, and what you need rather than who you are.

It's not about collecting statistics and figures to attract the right audience; it's about finding people and what drives them — defining them, recognizing them, and approaching them with content directly directed at them.

## 2.6 Promotion of YouTube Channel

**Fan funding in YouTube channel**

Tim Schmoyer says a Patreon campaign's most important part is to have a dream of why you're trying to raise funds. You are going to want to explain what you want to do, why you need money, how the funds are going to be used, and how the Patreon system works. By using low-cost benefits that inspire people to contribute, you will want to make it easy for your viewers — set goals for fans to be excited.

**How to Use Tubestart for Crowdfunding Support**

Dane Golden says some of the features of Kickstarter / Inidegogo and Patreon are available from Tubestart. The website offers four different funding methods, so you can choose the one that is right for you:

**Subscription Financing:**

Subscription Financing is the most similar option to Patreon from Tubestart. Customers can apply to an ongoing monthly subscription program where they purchase ongoing perks as long as they pay, or they can purchase one-off perks, and the developer is charged for the transaction. The subscription fee for Tubestart is 4%.

**Pledge funding:**

Pledge funding is for an ongoing project where backers pledge to pay a certain amount each time a new video is released by creators, or they can buy one-off benefits, and creators are paid on that transaction. Tubestart's pledge funding fee is 4%.

**Fixed funding:**

Fixed funding is for a one-time 30-90-day project where creators only receive paid contributions if they meet or exceed their contribution objective. Tubestart's fixed funding rate is 4%.

**Flexible funding:**

Flexible funding is for a one-time 30-90-day project where creators get paid all the money raised as it comes in, even if they fall short of their target amount. Tubestart's flexible funding fee is 4% of creators achieve their goal and 8% if not. Tubestart charges initially 8 percent for all purchases with this project, and then if developers reach their goals, Tubestart credits the maker with the additional 4 percent.

## YouTube Fan Funding ("Tip Jar") For Your Channel

YouTube also started rolling out Fan Funding, which is incorporated with their program, also known as the Tip Jar. It helps fans to make a one-time donation to the makers. YouTube pays twenty-one cents for each payment plus a 5 percent fee, which goes into Google Wallet. You can't contribute to it often. This system is rolled out through many countries slowly.

# Chapter 3: Monetizing your YouTube channel

You've learned the algorithm for YouTube. The SEO of your channel is in tip-top form. You also decoded the metrics for YouTube. Finally comes the moment that you were waiting for — it's time to make some money to do what you do.

Firstly, the channel must have at least 4,000 hours of watch time in the last year and at least 1,000 subscribers to qualify for monetization. This rule came into effect in early 2018 and is another way for YouTube to prioritize watch time (as opposed to view count, which was the previous emphasis of the platform). "Watch time" is the average number of hours spent by users watching videos from your website. Because YouTube mostly wants to keep viewers as long as possible on the site, viewing time has become increasingly valuable to creators.

The 1,000 subscribers/4,000 hours of watch time benchmarks are pretty realistic based on conversations in the r / YouTubers and r / YouTube subreddits! Plus, it won't be long before you hit the magic numbers once you've implemented an active SEO strategy.

Don't try to play the game. If your channel shows any signs of participating in "sub4sub" or "view4view" tactics, your request may be rejected. In other words, if you subscribe to other people's channels in exchange for subscriptions to your channel, or watch videos from other channels in exchange for watching hours on your channel, YouTube can find out your monetization and refuse it (or disable it). Don't do it that way.

YouTube, just like google is a search engine, so there are ways to improve your exposure, but the sophisticated algorithm of YouTube will detect it and penalize your channel if you do anything to play the system. Committing to the shady tactics of YouTube is a risky move and a waste of time. Trust your artistic instincts and focus on producing content you enjoy and get the subscribers you deserve.

**Prepare your Monetization Platform**

You can work on some things to ensure that the channel has the best chance of monetization before you hit the "Enable" button to monetize your channel. It can also take up to 30 days or more for YouTube to review your channel, and if your application is denied, you have to wait another 30 days to apply, so do not take any chances — you want revenue to start coming in as soon as possible.

Copying is the most common reason YouTube does not accept applications for monetization. (By the way, if your request is rejected, on your monetization site, YouTube will justify, but the reasons may be a little vague). Let's talk about how this issue can be avoided.

Only make sure you have all the correct licenses in place when using stock footage and audio. This is simply because every purchase from Story blocks of stock footage and sound comes with a license you can use at any time. Where to have and understand these licenses, even if you don't monetize your channel, is essential to know.

**Steps to allow YouTube monetization**

You have gained a number of required watched hours and checked for red flags on your channel— now what? It's time to learn how to enable YouTube monetization.

**1.** Drop-down My Channel click on the icon in the top right corner of the screen

**2** From this screen, click YouTube Studio (beta)

**3** Once you are on YouTube studio, find menu channel on the left side of your screen and click Other Features > Monetization.

**4** Finally, click Start in the Monetarisation window.

After this has been finished, you have accepted the terms of the YouTube Partner Program. Sign up for AdSense or connect to an existing account. This is basically what makes it possible for your network to make money. YouTube then prompts you to set your preferences for monetization. Don't worry too much about this step— you can later change your preferences for monetization and customize it to monetize only specific videos.

## Wait for Monetization Permission Sadly

It is not very easy to monetize on YouTube as clicking a button. YouTube typically takes 30 days to review an application. Occasionally, however, a backlog builds up. This is important to realize, as it means that for at least 30 days, you will probably not be able to start generating revenue.

Throughout this waiting time, all you can do is keep hustling. Keep posting regularly (the YouTube algorithm consistently rewards channels posting), make sure you're still focusing on SEO, and keep engaging with your audience. In the long run, if you cause your channel to lag, it could affect you. In contrast to monetizing, your channel can always take full advantage of additional subscriptions and watch time— both of which can open up new streams of revenue.

## 3.1 Four ways to monetize your channel

YouTube monetization is just one way you can make money on YouTube. There are several others, but it's a high, relatively easy entry point to enable YouTube monetization. The subscriber and watch time threshold for YouTube ads is pretty low so that when it's still in its infancy, your channel can pick up steam. However, as you continue to gain more followers and increase watch times, additional monetization options should be considered — especially if you have a very loyal and engaged audience.

## 1. Patronage

Patreon is one of the YouTube creators' most popular online patronage sites. Patreon allows your loyal followers to pay for access to exclusive content a certain amount of money each month. You can have multiple levels of patrons, supplying them with anything from behind-the-scenes footage to entirely new material (which cannot be seen by non-paying members). Many YouTubers give their patrons a day early access to videos; others even put the creators themselves in some one-on-one time. Patreon is an excellent fan-based choice for artists. This loyalty is gained more often than not by consistent communication and engagement with your followers.

## 2. Merchandise

Merchandise loves people. But before you start printing 1,000 t-shirts for all 1,000 subscribers, try the waters a little bit. Ask your audience if they want to buy goods from you. You can suggest merchandise types and even ask them to comment on the products they would like to purchase. Selling merch is particularly fantastic if you and your followers use specific jargon, catchphrases, or jokes inside. You can also include it in competitions, giveaways, or even add it to one of your Patreon rates when you start selling merchandise.

## 3. Affiliates

The good news is that the affiliate program is pretty straightforward. The earnings are just lower than what a paid sponsorship would give you. Once you register with a specific brand for an affiliate program, you will receive a unique discount code that your followers can use while shopping with that brand. You will earn a small commission every time you use your system. Mention the affiliate software in your videos and include it to boost earnings in your video descriptions, as well.

## 4. Service

Features Product features are a big step in the direction of paying sponsorships, but they may not result in money. Brands may send you their products to feature and review in your videos as you begin to gain a more extensive follow-up. It's excellent free stuff. You don't even have to wait until brands meet you — go ahead and pitch to potential sponsors. Only make sure you throw to businesses that make sense for your platform (although you may start receiving requests for random product features along the way).

## 5. Paid sponsorships

They are the gold mine on YouTube to make money. This is when companies are paying you in your videos to reference or support their products. We will be honest to you— this is a daunting goal to achieve. Yet preparing to sell yourself to businesses is never too early. Join them on social media, show them how to support their company, and offer examples of successful product sponsorships or affiliate programs you've done in the past. Even if you haven't hit the desired number of subscribers yet, you're still going to put yourself on the radars of brands. Again, make sure the companies you're looking for making sense of the content you're producing.

# Chapter 4: Types of YouTube Advertisements

Choosing the right type of YouTube ad needs gen and experience to achieve maximum returns from the investment, as the social media test and discovery of some form of YouTube ads becomes redundant, and new types are introduced.

## 4.1 True View Advertisement

True-View Advertising True-View advertising is the skip-able ads that appear at the start of videos from YouTube. For a few reasons, they are a great place to start advertising on YouTube. They are flexible True-View ads that allow you to advertise your products and services with clips, samples, video testimonials, and more.

Super dry, a UK-based outerwear company, received a "YouTube Works for Brands" award in September last year for their True-View advertising, "This is the Jacket." The campaign was extremely successful, resulting in an increase of 37 percent in digital sales and an increase of 55 percent in-hero product sales!

Effective True-View advertisements include a simple call to action from the beginning of the video, be it in the video itself, or through overlay advertising and supported cards (more on those later). We are one of the best ways to drive up to 500 percent of immediate customer engagement elsewhere on your YouTube channel!

Advertising from TrueView is a cost-effective way to reach a reasonable audience. You only pay if the ad has: played for 30 seconds or longer (or ended) Prompted the attention of the viewer (like a click) because TrueView ads can be skipped after 5 seconds, you cannot waste ad dollars on a totally uninterested audience.

True View marketing has a wider audience. When delivering advertisements, YouTube takes Google's search history into account, and you basically combine the audiences of the web's two largest search engines. Further, trueview advertising can be shown on other Display Network publisher sites depending on the type of trueview ad you choose: In-Stream or Discovery.

**Types of True View Advertisements**

**True-View In-Stream ads**

During YouTube videos, such ads play, and they can also appear in other places in Google's display network, such as applications or games. The in-stream ad may last up to three minutes, but it is generally recommended for 30 seconds. For your ad copy, there is no word limit, and clicks go to your website or digital storefront.

A companion banner— an actual show ad — appears in the top right as your in-stream ad plays. This is a clear route to your page.

**True-View Discovery ads**

Discovery ads are like display ads — in reality, in-display ads were even named. These are the suggested videos that appear on the website of YouTube or on the search page as recommended/related videos.

The length of the video with the discovery ads is not limited, because people choose to move through them. The corresponding show ad headline has a cap of 25 characters and can contain two lines in the body copy, each with a total of 35 letters. Generally, promotions for True-View are low risk, high reward. Even a missed ad will increase engagement!

Because of the option, up to 76 percent of customers reflectively skip ads. Nonetheless, it is still ten times more likely than people who Miss True-View ads would visit or switch to a product channel than those who have never been exposed to the ad. So, people who watch the ads, in fact? 23. The chance is 23 times greater!

**True-View for Reach**

TrueView for Reach, a new way to customize TrueView ads based on campaign targets, was launched by Google in April 2018. Instead of charging when a prospect watches the ad to completion, TrueView for Reach allows the pricing of CPM (cost per 1,000) for these shorter ads, which means you can pay for 1,000 views. To qualify for this type of optimization, your ad must be between six and 30 seconds.

If you want to reach a wide audience quickly, True-View for Reach is a good option. On Google's site, Samsung commented that they used the new format during beta testing to reach 50 percent by half their usual CPM.

Samsung, however, is a large company with a wide audience and product range, but when you advertise a more niche product or service, True-View for Reach may not be as effective, simply because it is possible that your marketing dollars will be better spent on your particular audience.

## 4.2 Non-Skippable YouTube ads

Advertisements that cannot be missed or maybe irritating, but they're here to stay. The good news is that people learn to tolerate them (or at least learn to do something else while they are playing), particularly when YouTube has reduced the maximum length from 30 seconds to 20.

### Types of Non-Skippable YouTube ads

There are two types of YouTube ads that cannot be skipped: pre-roll ads that appear before a clip plays, and Mid-roll ads that appear at the 10-minute or longer midpoint

You may want to set up some non-skippable YouTube advertisements if you're trying to tell a deeper, more complex tale that takes a little build-up. Make sure your video is fully focused and shows clearly to your target audience the value of your product.

Moreover, make sure the message is conveyed both audibly and visually by the video. And, even if somebody isn't interested in watching, they might still hear what you've got to say. The key to finding success with non-skippable advertising is skilful ad targeting. Take the time to refine the audience you like.

Non-skippable advertisements are generally effective in increasing exposure of the product to specific target markets. Considering, on a cost-per-mile (CPM) basis, they are paid for, allowing you more leverage over your ad spending.

## 4.3 Bumper Ads

Bumper ads are the variant of non-skippable ads that is more tolerable, lasting at most six seconds. At the end of YouTube videos, they appear and are paid for on a CPM basis. Because they are small, it is ideal for targeting mobile users with bumper ads. These are also a great way for longer content to be recycled.

In producing a six-second bumper model, cut down on the "annoying" aspect of non-skippable commercials. Then re-market to audiences who saw the version that could not be skipped to the shorter version. This way, you can improve brand exposure without upsetting viewers.

YouTube has a leader-board with the top 20-moment bumper ads based on views, click-through speed, and innovative user scores if you need inspiration.

## 4.4 Display ads

Such YouTube ad forms appear above the list of suggestions for videos. Whenever a user searches for a particular topic in the search bar of the YouTube page, the videos shown along with the display ads are recommended. You can also see display ads on the right side of the featured video occasionally. Such types of YouTube ads are rendered as a screenshot that clearly displays important information. The screen ad width is 300X250 pixels and 300X60 pixels.

Display ads are available in all sections, such as at the top of recommended clips, on the right of the video featured, between the playlist, except for the YouTube website or app homepage. Only when you play videos on YouTube on your computer can you see display ads. There is no mobile phone show of these types of YouTube ads.

## 4.5 Overlay ads:

Overlay ads are the sort of YouTube ads that can be seen at the video player's bottom or at the video's top right. Such advertisements are also referred to as "In-video advertising."

These are kind of semi-transparent banners that cover the video player screen's 20 percent. Often, overlay videos are only shown on the screen. Overlay ads contain 468X60 pixels or 728X90 pixel picture or text. Such advertising can be powered by Google's Ad Words. The maximum range of overlay advertising is open.

## 4.6 Sponsored cards

Sponsored cards are pop-up advertising designed to show between the videos on the video player monitor. Such types of YouTube ads are also cost-effective for businesses because they have to pay per received click or amount of overall views. The width of the sponsored cards for desktops is 300X250 pixels and is in the form of an image or flashcards, and the size of the sponsored cards is adjustable and for mobile phones in the form of jpg or jpeg or GIF (without animation).

## 4.7 Masthead format

Masthead is the most popular YouTube device or website advertising. Masthead marketing is like endorsed cards, but it occupies a large area on the website's home page and has maximum visibility. Masthead advertisements are ideal for large-budget companies. The resolution of the masthead home page is normally 970X250 pixels, but it can only be viewed on desktops. Furthermore, 970X500 pixel expandable mastheads can be shown for 24 hours at the top of the

YouTube website home page. There is another masthead card class known as masthead lite. The size of the masthead lite on the side of the YouTube video is reduced to 760X150 pixels along with the 265X150 pixel banner.

# Chapter 5: Maintaining Your Channel

## 5.1 Posting material multiple times a week

Recent reports have shown that channels on YouTube that post more than once a week perform much better and get more recommended views. Post a video on YouTube three or more times a week, if possible, particularly if you're just beginning and trying to build an audience. You will quickly raise your channel in the algorithm by maintaining a regular schedule of multiple posts per week.

Creating a ton of content on similar topics at the beginning will help your channel to perform well in the algorithm, as well as creating a content library that will use viewers from one video to another, boost your watch time and give them a reason to subscribe.

## 5.2 Develop Sustainable Video Work Flow

You may be producing Oscar-worthy short films and videos, but if it takes six months for each video to produce, your videos do not expand your YouTube channel. Regular video uploads are what brings people back for more at familiar times.

Whatever kind of videos you choose to make, select content you can frequently build and create and find ways to streamline your production workflow, whether it's setting up a studio, developing an editing template, recruiting assistants, or a production team. Keep your topics and workflow refined until a well-oiled machine is your process.

Setting up a small video studio is one of the best things you can do because when it's time to shoot, you can just switch on the lights and start.

## 5.3 Adding Interesting Hook

It depends on you and your material on how you attract viewers. If a video has any kind of plan, first reveal the end result. A stunning result will make people more interested in seeing how you got it. For DIY and makeover videos, this approach is great. Cute girls' hairstyles, for example, always start by showing the end result before explaining how to create a hairstyle.

Stories are another way of attracting the interest of viewers. For stories, people are hardwired. When you start a story with your video, people will naturally want to stick around to see what's going on. However, personal stories endear the viewer to the presenter and can often provide more complicated ideas with a helpful section.

To show, VSauce's Michael is a master at the beginning of his videos with an intriguing story or idea that leads to the subject.

You hook viewers; however, make sure that your opener is directly related to the subject. Viewers clicked because they were interested, so get quickly into the topic they first wanted to hear about.

You may have seen YouTubers create suspense by beginning with an entirely off-topic story or reality and then linking it to the topic. If you already have a large audience that trusts you, this tactic works best.

## 5.4 Keeping Title and Openings Short

There are short periods of focus. At the beginning of a video, a long title or credit sequence may cause people to lose interest. A long opener also discourages binge-watching because people do not want to watch over and over the same long sequence.

Instead, make short and punchy your opening title and credits. Allow no more than 5 seconds for the entire opener. The title sequence of Crystal Joy is short, sweet, and delightfully quirky as a great example.

## 5.5 Adding End Screens

End screens are interactive graphics that connect to your channel to another video, playlist, stream, or web page or prompt someone to subscribe. You can only insert end screens in the last 20 seconds of your clip, as the name of the function implies, so you need to prepare where the end screens will appear.

One choice is to position your video topic in a manner that gives you space for end screens. The audience will continue to talk about the end screen.

End screens tend to work better if there is still talk and information given to the viewer by the on-screen presenter. If you just cut a colour or design display and don't have any new information, viewers will probably turn off the page. YouTube viewers are conditioned to do this now somewhat. If you continue to provide information, it will give viewers a reason to stick around.

## 5.6 Edit Distractions from Videos

Long pauses, meandering talk, bouncing from one topic to another, or just being boring, can make people start looking for something more interesting in the recommended videos. Keep tangents to a minimum, and make sure that it engages either visually or with a story if you differ from the subject. Don't give a reason to move away to viewers.

You can quickly cut from one shot to another like Devinsupertramp to avoid distractions at the same time and keep your videos fast and engaging. You don't necessarily have to film with multiple cameras to create cuts. You can use text or transitions with basic video editors, as well.

## 5.7 Thumbnails to click on

Suggested videos are YouTube's leading source of organic traffic. Your clip thumbnail needs to stand out when it's a recommended video in the right sidebar as someone watches a video on YouTube. If your video appears as a recommended clip, YouTube essentially endorses it by suggesting you might enjoy your video as well as someone watching some other video.

Moreover, if your video attracts clicks as a suggested video, its click ability is likely to register with the YouTube algorithm. Remember that YouTube wants viewers to click on another video, more than anything else in the world. And they'll suggest the most likely videos to get the click.

First and foremost, make the thumbnail important to the title and content of the video. Nothing makes people click away faster than a thumbnail-free video. Spectators are feeling tricked. Not only will you alienate your viewers, but you will also have low viewing time, which is mostly rated by the YouTube algorithm.

Make viewers wonder what's next with your thumbnails; try telling a story. Show a picture that sets up a situation or teases it. Make the viewer wonder what's going to happen next or before. With the words "I Quit," Amy Schmittauer thumbnail tells a story that complements the title of the video.

Remember, thumbnails are just half the story. The second half is the title. A clever juxtaposition between title and thumbnail can go a long way towards increasing the interest of the viewer.

## Design Small

Many people create video thumbnails of 1280x 720 pixels, as recommended by YouTube. The pictures look great, but no one can ever see a thumbnail on YouTube at that size. While you want the pixel size to meet the guideline from YouTube, you still need to model the picture for a smaller viewing size.

To check how your thumbnail will really look on YouTube, always zoom out, so you view the image on-screen at the size that it will appear on YouTube. You want to make sure that the thumbnail image is still meaningful and stands out when a postage stamp is the size.

If not, it might be helpful to edit a few basic images. Try to crop the image into a smaller area or over-saturate the colours, as in the example here. The Sharpen tool can also define your image's edges. These edits may look bad at a large size, but your image will stand out at the size viewers will see.

## Create and look Consistent

If all of your thumbnails look consistent, you can see your videos at a glance. It may include a similar text font, a logo, familiar colours, shape or design element, or the face of the same person. Whatever you choose, on all your thumbnails, find something consistent and stick with the style.

**Be Emotional**

Excitation is the emotion that most people are reacting to. This makes us want to know why they're excited when we see someone showing excitement and maybe join in so we'll be excited too. Nothing sells more pleasure than the eyes, so show an enthusiastic face and focus on the eyes. Your video is going to get a much better answer.

**Test Thumbnails with AdWords**

Ultimately, unless you test it, you won't know if a thumbnail is going to work. Create multiple options for Google AdWords to test your thumbnails.

Then spend $10 a day seeing which thumbnail gets the highest view-through rate (VTR) for about a week. Especially if you run a campaign or take advantage of a tentpole event, this tactic is worthwhile.

The Suggested Video views on YouTube Analytics can help you monitor your thumbnails ' success or progress. Open Creator Studio to group the videos you want to test and go to Analytics > Overview. Click Groups in the top right corner and from the drop-down menu select Create Video Group.

Next, select Analytics Traffic Sources and then Suggested Videos to see which video was best performed in your group.

## 5.8 Tactics from top-performing Videos

Nose-to - the-grindstone work is no alternative. Take a deep dive into your analytics to find out which videos get the highest number of subscribers per view. Look for patterns like subject, delivery, or editing style among them. Then make more such videos.

To find your Creator Studio's highest conversion clips, go to Analytics, select Subscribers, and then click YouTube Watch List.

If you want to get more accurate, a subscriber ratio can be calculated by dividing the number of subscribers by the total number of views per video. Be aware that, because most people subscribe to the channel page, the number will be really low.

## 5.9 Making Long Videos to improve watch time

While you want to keep your credits short, make the actual content of your video as long as it makes sense for your subject. It seems counterintuitive to make long videos, given the online spans of the famously short attention. In reality, shorter videos have been better considered. But longer videos today are more time to watch, which boosts your algorithm content.

Essentially, the optimal length for a video is long enough to relay all the data without padding the video. Don't make a video longer just because you're going to lose viewers to make it longer. But with longer videos in mind, you want to improve your video content. Generally, 7-15-minute videos tend to perform best.

PBS Space Time is a popular channel with lots of videos falling within the overall range of 7-15 minutes, but also having longer videos going nearly 20 minutes. You can even find channels that post videos for 30 minutes on a regular basis.

## 5.10 Going Live on YouTube

Live streaming is a great way to get content out of it without spending a ton of time on it. While live streaming has a definite learning curve, live video is the easiest way to create video content after you've mastered the format. On all social media platforms, live streaming is strongly supported. (The live video feature on YouTube is YouTube Live.) This video format offers excellent opportunities for engagement because you can communicate with your audience directly. Live video also has a long time to watch. You can use your Smartphone or webcam for live broadcasting to start sharing live video. Even webcams that are relatively cheap still can provide high-quality video.

## 5.11 Using Videos in Content Mix

You post videos that have one of three goals with the "Hub, Hero, and Help" strategy: serving the community around the channel, being shared, or performing in search. Those clips are added to your content calendar at different intervals: monthly uploading of hub videos, quarterly hero videos, and daily supporting videos.

Hub videos are focused on the community and are designed to create a strong bond between your viewers and you, your viewers, and each other. You may be responding to comments in hub videos, answering questions, interacting via live video, initiating projects that require the participation of the viewer, or sharing personal stories that give fans a glimpse of your channel behind the scenes.

Hero videos are videos of the tent pole built to be very effective. These videos are often focused on topical issues such as holidays and news events. Put extra work into a hero video to increase the likelihood that it will be posted by people on social media and/or forums and picked up by other media.

Help videos are designed to have content that is highly searchable, providing actionable value-creating clips such as DIY, how-to, tips, and troubleshooting. Focus on the search for discoverability for these.

Google Trends is an excellent resource for hot subjects and highly searched material to help pick topics for both heroes and support videos.

## 5.12 Make Videos in Series

Nothing rises, viewing time as watching binge. Basically, for your viewers to go smoothly from one of your videos to the next, you want to create a "lean back" experience. Series playlists are one of the best ways to do this, yet on YouTube, they are an underused tool. Nonetheless, you need a sequence to use them.

You may create a content-based series that has been successful. Yes, there are many popular YouTube channels with more than one episode, each with a particular focus on the subject and even different styles of thumbnails. To illustrate, for unboxing and tech reviews, Roberto Blake uses one thumbnail style and another for tutorials:

You can add videos to a series playlist when you post them in your series. YouTube automatically adds the next video in the playlist to the Up Next section at the top of the recommendations when you use a series playlist. And if a viewer clicks on auto play, the playlist clips will run one after the other.

One note about series playlists: although a video may be in as many playlists as you want (and it is recommended that each video be added to at least three playlists), only one series playlist may contain a video. So, make sure that your series playlists are high-level categories and that your regular playlists are more specific.

## 5.13 Collaboration with other You Tubers

Collaborations are videos shared by multiple content creators and one of the most effective ways of expanding your YouTube audience. Whether you're a brand or a YouTube, working with another producer of content will introduce your channel to people who probably never heard of you.

Social proof has a powerful influence on people. It is an acknowledgment from their audience when a content creator collaborates with another developer. Checking your channel can be enough for a viewer, and if they like your content, you have a new subscriber.

People often forget that YouTube is a social network in the first place. And by cross-promoting and collaborating together, some of the largest creators became successful. So, while allowing your competitive side to take over when you see a channel gaining more popularity or growing faster than yours may be easy, try to see their development as an opportunity for you to grow with them.

Yet you can't work with anybody. While looking into this strategy, you need to understand some finer points of cooperation in order to grow your network.

YouTube's most successful collaborations are with channels that have a similar audience but cover entirely different content from yours, as the audience will not get your subject matter repeated on their channel. The fact is all of us, including your audience, have multiple interests. So, it's a great opportunity for growth wherever you can find overlap.

For example, a channel for science education might share an audience with a channel for sci-fi movies. A gaming channel may share a tech review channel with an audience. A family channel could share a toy review channel with an audience, and so on.

Generally speaking, you want to find outlets that are slightly bigger than yours, but not too big. We are asked to partner with super-large networks all the time and usually only work with people we meet (so get to know them if you can!)

Working with a smaller channel can be great if it's a good fit and especially if it grows fast because the tiny YouTube of today could be the big breakout star of tomorrow.

Collaborations don't just have to be deals between two people. For videos, multiple creators may gain exposure to different viewers. Tyler Oakley often collaborates with other creators of different genres and, while developing relationships in the process, has grown his audience exponentially.

The will your network expands, the more partnerships you create, the more opportunities you will have to interact with multiple creators. Many of the largest YouTube channels do this all the time, simply because they become friends with these other creators and hang out in real life together. Indeed, a rising tide lifts all boats.

Connections with other creators make YouTube the social media site it is, and by collaborating with multiple creators, you can build tight bonds and teams that work together to go much further.

# Chapter 6: SEO (Search Engine Optimization) on YouTube Videos

Search Engine Optimization (SEO) is the ever-changing method of web content development that will rank high on search engine results pages (SERPs). Because search is often the gatekeeper of your content, it is important to optimize your website for search to attract traffic and develop a search.

YouTube SEO involves the optimization of your own site, playlists, Meta information, definition, and videos. To search both inside and outside YouTube, you can optimize your videos.

## 6.1 YouTube Keyword Search

YouTube's SEO process starts with work on video keywords. Here is how to find the right keywords for your YouTube videos: generate a list of ideas for keywords first. Just as you'd like to use an SEO tool first to identify keywords that you want your video to focus on when optimizing written content (you can browse popular YouTube SEO tools below these tips, or just click that link earlier in this phrase).

Before you even upload it to YouTube, the first place you can bring it is your video file. YouTube cannot actually "play" your video to see how important it is to your target keyword, and as you will find in the tips below, there are only so many places where you can securely add this keyword on the preview page of your video once it has been released. But when it's uploaded, YouTube will read the file name of your video and all the code that comes with it.

With that in mind, replace your desired keyword with the file name "business ad003FINAL.mov" (don't be embarrassed... we were all there during post-production). For example, if your keyword is "house painting tips," the file name of your video should be "house painting tips" followed by your favourite form of video file (MOV, MP4, and WMV are some of the most common ones that are YouTube compatible).

## 6.2 Inserting your Keyword Naturally

One of the first items that our eyes are drawn to when we browse for videos is the name. Often that's what dictates if the audience chooses to watch the video or not, so the name should be not only convincing but also clear and concise. With that in mind, replace your desired keyword with the file name "business ad003FINAL.mov" (don't be embarrassed... we were all there during post-production). For example, if your keyword is "house painting tips," the file name of your video should be "house painting tips" followed by your favourite form of video file (MOV, MP4, and WMV are some of the most common ones that are YouTube compatible).

While your keyword plays a big part in the title of your video, it also helps if the title fits closely what the viewer is looking for.

Backlinko's research found that clips that fit the exact keyword in the title only have a slight advantage over those that don't. Here is a linear representation of the findings:

So, while "using your target keyword in your title can help you rank for that term," says report author Brian Dean, "the relationship between keyword-rich video titles and rankings" is not always strong. Nonetheless, adopting the title for this keyword is a good idea as long as it actually blends into a name that tells viewers exactly what they're going to see.

Also, be sure to keep the name relatively brief — Alicia Collins, HubSpot campaign manager, suggests keeping it to 60 characters to help keep it from being cut off in results reports.

## 6.3 Optimize the definition of your video.

First things first: The maximum character cap for descriptions of YouTube videos is 1,000 characters, according to Google. And while using all that space is all right, remember that your viewer most likely came here to watch a video, not to read an essay.

If you choose to write a longer summary, note that YouTube shows only the first two or three lines of text— around 100 characters. After that point, to see the full description, viewers need to click "show more." That's why we suggest that the description be front-loaded with the most important information, such as CTAs or key links.

It does not hurt to add a transcript of the video to improve the video itself, particularly for those who need to watch it without size. That said, the research conducted by Backlinko also found no correlation between descriptions optimized for a specific keyword and the rankings for that term.

Nevertheless, Dean is cautious not to promote the full dissection of an engineered definition. "An integrated summary makes you appear in the suggested sidebar of the images," he says, "which for most channels is an important source of views."

## 6.4 Tagging Video with Proper Keywords

The official Creator Academy of YouTube recommends that you use tags to let viewers know about your video. But not only do you inform your viewers — you inform YouTube itself as well. Dean states that tags are used by the platform "to understand your video's content and meaning."

YouTube will thus find out how to combine your video with similar videos, which can extend the reach of your content. Yet carefully pick your marks. Do not use an insignificant tag because you think it will give you more views— in fact; you might be penalized for that by Google. And similar to your description, lead with the most important keywords, including a good mix of the common ones and the long-tail ones.

## 6.5 Categorizing Your Video Content

Once you upload a video, you can categorize it under "Advanced settings." Choosing a category is another way to group your video with similar content on YouTube, so it ends up in different playlists and gains exposure to more viewers who identify with your audience.

Maybe it's not as simple as it looks. Nonetheless, the Creator Academy of YouTube recommends that advertisers undergo a rigorous process to determine which class that video belongs to. It's beneficial, the guide says, "to think about what fits well for can class,"

## 6.6 Customizing your Thumbnail for Video

When scrolling through a list of video results, the video thumbnail is the main image viewers see. The thumbnail sends a signal to the audience about the content of the video along with the name of the video, so it can affect the number of clicks the video gets and views.

While you can always choose one of the auto-generated YouTube thumbnail options, it is highly recommended that you upload a custom thumbnail. The Creator Academy estimates that "90 percent of YouTube's best-performing videos have custom thumbnails," encouraging the use of 1280x720 pixel images — showing a 16:9 ratio— that are saved as 2 MB or smaller.jpg,.gif,.bmp, or.png files. When you obey these criteria, it can help ensure that your thumbnail appears on various display channels with the same high quality.

It is important to note that to upload a custom thumbnail image; your YouTube account must be checked. Visit youtube.com/check and follow the instructions mentioned there to do this.

## 6.7 Using SRT Files to Add subtitles & closed captions

Subtitles and closed captions will improve YouTube search efficiency by highlighting essential keywords, like much of the other text we've covered here.

You will need to upload a licensed text transcript or timed subtitles file to add subtitles or closed captions to your video. For the former, in order to auto-sync with the video, you can also explicitly insert transcript text for a clip.

Adding subtitles follows a similar method, but the amount of text you want to view can be reduced. For either of these, go to your video manager and click "Videos" under "Video Manager." Find the video to which you want to add subtitles or closed captioning, and then press the drop-down arrow next to the edit button. Then choose "Subtitles / CC." Then choose how to add subtitles or closed captioning.

## 6.8 Adding Cards & End Screens to increase your YouTube Viewership

Have you ever seen a small white, circular icon with an I in the centre appear in the corner when you watch a video or a translucent text bar that asks you to sign up? These are Cards that are described by Creator Academy as "preformatted notifications that appear on desktop and mobile that you can set up to promote your brand and other videos on your channel."

Up to five cards can be added to a single video, and six types are available: channel cards that direct viewers to another channel.

**Types of Cards**

Donation cards on behalf of U.S. non-profit organizations to promote fundraising.

Fan funding is available to ask your fans to help create your video content.

Link cards that direct viewers to an external page approved crowdfunding platform or an approved platform for merchandise sales.

Poll cards that ask viewers a question and allow them to vote for an answer.

Video or playlist cards that link to this form of other YouTube content.

End screens display similar data to cards, but they do not show until a video is over, as you may have expected, and are, in fact, a little more visually complex. A good example is the overlay with a book image and a visual link to see more in the following video.

There are a range of detailed instructions on how to add end screens depending on what kind of channel you want to model them for, as well as different types of content that YouTube requires. Google explains the specifics of how all these aspects can be implemented here.

It is important to note that YouTube is always reviewing end screens to try to improve the viewer experience, so there are occasions when "the end screen may not appear as you have selected it." Take these factors into consideration as you choose to use either cards or end screens.

Such variables may seem a little confusing and time-consuming but remember: over the year, the time people spend watching YouTube on their television has more than tripled.

Now, most of the above SEO tips rely on you to correctly identify a keyword and promote your image. And not all these tips can be done by YouTube alone. Find some of the tools below to automate your video search to get the most bangs for your video buck.

## 6.9 YouTube SEO Tools

**Ahrefs Keywords Explorer**

Ahrefs is a comprehensive SEO platform that enables you to monitor the ranking of a website, estimate the organic traffic you would get from each keyword, and identify keywords for which you may want to create new content.

Keywords Explorer has a popular feature, which allows you to look up to various information related to a keyword. And as you can see in the screenshot above, search engines, like YouTube, will filter your keyword results.

Ahrefs Keywords Explorer shows you the monthly search volume of a keyword, how many clicks for that keyword, similar keywords, and more earned through video ratings.

**Canva**

Canva could be known as a template of design to create all kinds of cards, photos, logos, and more. This popular product also happens to have a Thumbnail Creator for YouTube videos.

As described in the tips above, thumbnail images are essential to promoting the search results of your content on YouTube and encouraging users to click on your video. You can create the perfect preview image for your video in 1280x 720 pixels using Canva's Thumbnail Creator— the thumbnail dimensions required by YouTube.

**Hub Spot Content Strategy**

Our content management tool, built here at HubSpot, allows you to find common keywords to create content and then organize these keywords into groupings — what we call "topic clusters." By sorting your content into topic clusters, you can track which content pieces are connected to each other, what content types you've designed, and what you've produced.

Clustering your content — and linking videos to blog posts, and vice versa — will give you more authority in Google's and YouTube's eyes, while giving you more ways to capture traffic from people looking for your subject.

**Vid IQ Vision**

This is a Chrome extension available in the link above through the Chrome web store, which lets you examine how and why those YouTube videos are performing so well. Then the vidIQ tool offers an SEO "rating" that you can use to create content that will match (or outperform) the results you see on YouTube already.

**Tube Buddy**

TubeBuddy is an all-in-one video platform that helps you manage your YouTube content's production, optimization, and promotion. Its features include a non-English keywords automatic language translator, a keyword explorer, tag suggestions, a rank tracker for your published videos, and more.

**Cyfe**

Cyfe is a large suite of software that offers a platform for web analytics, among other things. On this platform, you can track the performance of each website property on which you have content — including YouTube — and where the traffic on each page comes from.

Cyfe can also show you, in addition to traffic analytics, what keywords you rank for and which are most common across different search engines. Google Analytics or Moz sounds a lot, right? That's because Cyfe has built-in information from both these tools and more.

# Chapter 7: Best YouTube channel Ideas

Around the world, YouTube has more than a billion users. So, if you want to build a following, having a YouTube channel is important. However, with so many video formats out there, understanding what video concept of YouTube is best for your brand can be a challenge.

Well, don't fight anymore! We have collected the most innovative video ideas we could find on YouTube — from channel partnerships to gaming videos. So, get ready to tour through ideas that will revitalize your brand and give more to your viewers. And make sure that our free mood board template captures the good YouTube ideas as you go. Let's go for a dive!

**Personal Story Video**

You will need to create a channel that is somewhat special to get noticed on YouTube. Share a personal story in order to show the world what makes your YouTube channel special. Personal stories can help viewers connect with you and give them an insight into your life. If you have to tell some very interesting stories, they'll want to find out more, of course. Make a fun channel trailer that introduces viewers to the type of content that you will include in your stream.

**Product Review Idea**

Show and post your own personal analysis of a newly released item on your site. Try to distinguish the review with insightful commentary and thorough editing from others. The brand you select is on you, but if it falls into your area of expertise, you're likely to be able to make a very informative video about it. Tech reviews are already very popular video ideas for YouTube, and if you're sharing a fresh perspective on a new piece of technology, your video could attract a lot of attention.

**Comedy Sketch Video**

Laughing is a surefire way to boost your subscribers. That's more easily said than done, though. Comedy is subjective, so what makes a person roll around the floor in laughter may not even make others smile. Play your strengths and create a trustworthy sketch to perform. Be yourself and let it shine through your own unique sense of humor. You never know, there may be so many kindred spirits out there that will love your video ideas on YouTube.

**Lists**

The interesting list is always a magnet for YouTube users, so viewers are more likely to stay until the end if it's well described. These video ideas for YouTube are often structured as top 5, top 10, or even top 50 lists for some of the favorite topics of a YouTuber. Provide a thorough explanation of why you somehow organized the list. Choose a subject that you can easily explore and share with the world your passion and views.

## Unboxing

The often-deduced unboxing video format remains immensely popular among viewers of YouTube. Unboxing videos are a favorite among kids who want to check out the latest toys, but with other age groups they are also a big hit. For a younger audience, channels like Ryan Toys Review are appropriate, whereas channels like Unbox Therapy are suitable for a more mature audience. Choose a niche you know when you start unboxing and remember that irrespective of how well your video is presented, your audience must be concerned about the product first and foremost.

## Tutorials

Tutorials are one of the most growing video ideas for YouTube. If you can provide somebody with experience to help them accomplish something or complete a mission, they will be very interested to watch your videos. Nearly any subject can be covered by tutorial videos. Creating tutorials, gaming tutorials, cooking tutorials, language tutorials, and engineering tutorials are common forms of tutorials. Tell your viewers how the skills you're teaching them can be used. When you send a makeup tutorial, for example, show how and where to use that look.

## V-logs

Vlogs give YouTubers the chance to talk about their lives and the things they're most interested in. Getting off the ground may be one of the toughest YouTube video concepts, but if you manage it, it's worth the rewards. Content is king in vlogging, so keeping yours interesting and up-to-date is important. Also, popular vloggers have started from the very bottom, like Casey Neistat, so don't be put off by the challenge ahead.

## Challenges

In recent years, there have been a lot of clips with different challenges. Sometimes these videos can capture public attention from the ice bucket challenge to the cinnamon challenge on a phenomenal scale. Make a video of your own challenge and add a twist. Do the challenge to create more interest in a costume or a special venue. The most famous are often difficult videos that go wrong but do not deliberately undermine your efforts.

## Pranks

For several years, pranks have been one of the most successful video ideas on YouTube. Many strong-profile Youtube stars, like Roman Atwood and Vitaly Zdorovetskiy, focused all of their channels on performing outrageous pranks. Some of these pranks are large-scale and are not practical for beginners, but they can be as amusing as smaller pranks. Your friends and family are the easiest targets, as they will most likely forgive you when they realize that they were pranked!

## Parodies

Find celebrities, politicians, movies, or songs people know about and make a parody of them. If your subject is well known to your audience, it will strike a chord. Plan the scenario and get your friends on board to help you out. Until you start, it's important to get a good idea of what you want to film. Try to stay short and to the point of your parody. If your video runs too long, viewers who don't know your channel can be put off watching.

## Interviews

Keep an eye on your area's media and look for your next chance to interview people. You don't have to be popular for your interviewee; if they have an interesting story to tell, audiences will want to hear about it. When you can show the human side of a story that anyone can relate to, the video content would be even more appealing.

## Response Video

YouTube once had a standard video reaction feature, but its failure was due to low levels of usage. That doesn't mean you can't get a video response, though. Be sure to include the exact phrase of such a video in your own title when responding to a popular YouTube video. It makes it clear that the two are connected to whoever finds your picture.

## Tours

Room or house tours are sometimes seen by viewers as a common form of YouTube video. Tours give you a glimpse of your private life and an opportunity to show you what you're passionate about. Tours are one of the best concepts for incorporating YouTube video. All you need is a camera and a place to get started. If you stay in your room or house, display items like furniture, tech, and collectibles. It features a great spot for a more detailed tour that may not be known to many people.

## Time-lapse

A time-lapse can reveal new information about a scene that a regular video wouldn't notice. These days, most cameras have a setting that can capture very easily a time-lapse. Nonetheless, if you need to manually record the time-lapse, you will need to consider variables such as the short duration and the distance between images carefully. Set your camera in a fascinating space where a lot of activity is taking place. If you manage to capture something that happens fascinating, viewers may be tempted to watch.

## Question and Answers

It shouldn't be your first YouTube video to make a Q&A clip, but it can be a very useful YouTube video idea to add after you have built up some interest in your channel. Ask your viewers at the end of your previous video for their questions and then pick your choices. By posting on Facebook, Instagram, or Twitter, you can also get some additional questions. Q&A could cover a variety of topics. You can answer questions based on your uploaded videos, or you can do one about your personal life.

**Rants**

A blog for YouTube isn't something you should think about long and hard. You'll be able to rant easily about the subject if you've been riled enough. Write a loose script if you're not confident about ad-lib. After all, making sure your rant makes sense is important. Include a few jokes in your rant to make your video amusing. This will help keep viewers engaged and may increase their chances of tuning in to your next rant.

**Animal Videos**

Animal videos that do funny or unusual things make up a significant portion of YouTube's viral content. The right candidates for such a video might be your animals. Start filming them randomly and see if they do something worthy of being featured on your channel (treatments also help to encourage them). Who knows, you might have a YouTube star on your hands if your animal clip catches on.

**Tips**

Similar to instructional videos and explanations, tips videos are intended to remove viewers ' details, but they are typically condensed and do not go into as much detail as possible. Tips clips can be described as '10 tips for improving X' or '20 tips for improving your Y.' They are often informal, casual, and humorous. You can make video tips about a game you've been playing, a place you've been, or a job you've completed.

## Discussion

To get people to talk on YouTube, begin a thought-provoking debate. Introduce an engaging topic and ask in the comment section for the opinions of people. Everyone has an opinion, so it may be easier than you think to attract viewers to watch your discussion, especially if it's based on something, they're very passionate about. If your video is good, other YouTube users may be able to spark reaction videos. This can result in a series of videos on the topic gaining more views on your page.

## Draw my Life

Drawing my life videos involves YouTubers sharing in the form of sketches, a simplified version of their life stories. The video will show you drawing the pictures on an empty yet fast-motion screen. You will share new information about yourself that you may not know about your subscribers. For some information on this type of YouTube video ideas, check out the Draw My Life page. Most videos of my life have received millions of views.

## Things to Do

YouTube users will be looking for new things to do when they get bored. If your channel can provide them with interesting ideas, their attention will be piqued. You can create a title that outlines this specifically for those who view your video, like' 10 things to do when you're bored.' Video-making items can also give people ideas on what they can do in specific situations or on vacations.

**Cute Video**

Cute videos are the most amazing video ideas on YouTube, and will always be popular with a wide variety of viewers. Catch your baby's sister or pet doing something cute and share it to watch your subscribers on YouTube. Instantly, videos of adorable kids or pets can make us smile, and they can be the Kickstarter for your YouTube channel.

**Life Hacks**

Sometimes life can be challenging, and your audience will appreciate anything that can help make it a little easier. YouTube is hugely popular with videos of lists of different life hacks, and many have tens of millions of views. You may have some life hacks you've been using for years, but you've never posted. Now is the best time to share it with audience.

**Tag Videos**

The YouTube video ideas group includes one channel tagging another and asking them to upload a particular video or complete a particular challenge. There are so many videos you can do with tags. Get your friend to challenge you without being questioned, or even just do a tag picture. Several possible video ideas for YouTube tags include tag of guilty pleasures, a tag of my love story, tag of mother, tag of birthday, and tag of makeup.

## Custom Animations

Custom animation is a fun way of telling a story. It's fun to engage with viewers and create an animation focused on something topical and/or humorous. Animations can be used for the company or event as promotional material as easily as they can be used to inform people about important issues. If you're a professional animator or illustrator, this is one of our YouTube video ideas to put into practice.

## Comparison

Comparison videos are useful to YouTube users who want to find out about rival product information before purchasing one of them. Give viewers a good insight into what every brand provides, while highlighting its advantages and disadvantages. In your video, you might compare two or multiple items. At the end of the video, the good idea is to give the products a rating that makes your opinion clear about them.

## Showcase your talent

If you have a wonderful talent, skill, or ability, it could be a great idea for YouTube video to show that to all your subscribers. YouTube viewers are all very popular with dancing videos, soccer skills, and singing videos. Even though the talent is a little darker, audiences may still be interested in seeing it. In reality, having a highly unusual talent is likely to get even more mobile.

## Behind the Scenes

Go behind the scenes to show how you plan and send your videos to your subscribers. Highlight all the processes involved in making content of high quality and send them advice on how to make their own videos. To give your subscribers a good laugh, include bloopers and outtakes from some of your latest videos.

## Daily Routine

Let individuals understand what makes you tick by creating an everyday routine video. Daily routines will include eating habits, daily schedule, schedule of workouts, etc. We give viewers a bit more insight into who you are and how you are. Share the daily things you do that you know your quality of life is increasing and show your viewers how it can help them as well.

## Celebrity gossips videos

The world's obsession with celebrity culture won't end anytime soon, so there's plenty of material to make videos on YouTube of celebrity gossip. You may release daily videos that discuss and offer your own reaction to the latest celebrity gossip. To make your clips more memorable, send your viewers a special and funny take on the gossip.

**Business Intro**

Promote your business with a custom-made enterprise video on YouTube. Describe the type of business you are in detail and provide some background information. Display the company logo and corresponding tagline clearly in the video at some point so that viewers can identify with your product. Having your business video intro look professional is important, as the first impression that viewers get from it is the norm on which you are judged. Video Wizard has a range of great video templates for company that you can customize to create your own branded video.

**Customer Testimonial Videos**

One of the most popular YouTube video ideas for getting new customers is to add customer testimonials to your YouTube channel. Customer testimonials are often more reliable than self-promotion to attract new customers. They show that the company actually helped real people. If a potential customer is interested in your brand, they will try testimonials from the consumer to see what other customers got out of it.

**Product Launch Video**

Go live for the launch of your latest product on YouTube. Maximize the impact by pre-launching hype on all your social media platforms. YouTube Live is a perfect way to communicate in real-time with your viewers. Invite them to ask your product questions and then answer them after the launch. Keep in mind that the product can make a difference in the lives of people. What is the dilemma they are facing, and how is the answer for your product?

**Collaboration YouTube Idea**

In a YouTube joint video, Cross promotes with other companies that benefit both parties. You may have access to a completely new audience by starring in a clip of another company. If the audience is too different from your own, however, these YouTube video ideas like this one may not be successful. It's important to give your new audiences a good first impression as an opportunity to talk to them may not come back.

**Promote an Event**

YouTube is a great event promo website. That doesn't mean you ought to be overly promotional, though. Present the activity in a way that will make your audience feel something, and your video will be a great source of motivation. Search for clips of past events that will resonate with your viewers and help convince them that this time around, they can't miss it.

**Showing off Your App**

YouTube is a great website for promo activities. That doesn't say, however, that you should be unnecessarily promotional. Present the action in a way that will make your audience feel something and a great source of inspiration for your film. Search for clips of past events that resonate with your viewers and help them to convince them that they can't miss it this time around.

**Launching a Competition**

Use YouTube to launch a competition and invite viewers to share their social media image. Competitions are a great way to get your audience involved. Consider the award enticing, and more people are going to come in. You may create a new YouTube video to reveal the winner once the competition is over.

**Give Business Advice**

Giving a business advice can be a very good video idea for YouTube if your business has been around for quite some time. Share the secrets of your success with your audience and give them tips on how the methods that worked for you can be replicated. Such videos can help attract the type of business-oriented viewers you're hoping to engage with.

**Product Tutorial Ideas**

This might be the great way to advertise your product to demonstrate how it works to your audience. Make sure that you demonstrate the best bits and exactly how they work. Create a playlist and receive questions from your viewers with the rest of your tutorials. The favorite playlists for YouTube, so make sure as many playlists as possible. If the product you released is quite complex, you could create a series of tutorial videos that focus on specific product areas.

## Introduce your team videos

One of the more light-hearted business-related video ideas for YouTube is to give the company a look behind the scenes to your subscribers. The launch of your group gives you the opportunity to show your work culture, express your sense of humor, and increase awareness of the brand. It is also a related way to engage your audience, as they may see parallels between themselves and some of your staff.

## Beauty Make up tutorial

Share it with the world if you're a makeup whizzes and starts creating tutorials on YouTube. Display all the tips and tricks you've picked up over the years to your viewers and describe how your favorite looks created. Highlight (and contour!) the items that you use to produce the look that you present in the tutorial and what it costs to buy.

## Hair Tutorial

Hair tutorials are one of our YouTube video ideas that are close to makeup tutorials. You can either choose to do both or concentrate your channel theme on either. If you're serious about styling hair, this could be a path you'd like to try. Give your viewers tips on how easy and complex doses of hair can be copied. Check out some of YouTube's most popular hair tutorial videos and figure out ways to improve or do them yourself.

## Favorite Products

Most popular You Tubers are paid for using certain makeup products, but there is plenty of demand for videos showing what the real favorite products of people are. Take a dip in your makeup bag and show off the things you can always count on to make you look your best. Including foundations and concealers to brushes and mirrors, these could be anything. It's your choice!

## Clothes Tips

Do you have an odd knack to know what's going on with? If you do, share the enviable sixth sense on YouTube with viewers. Look at different types of styles or specialize in a particular style in different videos. For example, you might look for summer dress tips, clothes going out, winter wear, etc. Show off your favorite clothes and explain why you like them so much to everyone.

## Fashion and Beauty Trends

Look at the latest fashion and beauty trends and explore them in your YouTube videos. If you are able to provide a trusted source of fashion and beauty news, viewers will be more willing to subscribe to your channel. Give your opinion and don't be afraid to discuss these patterns with your mind.

Interestingly, not afraid of disagreeing with the status quo are the most popular YouTubers.

## Reaction to Your Old Makeup

Use them as a chance to get new viewers. While you may react at some of your previous looks, reaction videos may be funny YouTube video ideas and may draw a lot of views. Get few of friends and family to comment on your old makeup and put yourself in the firing line.

## Makeup Challenge

On your YouTube channel, take on and feature a popular makeup competition. Some makeup competition YouTube video suggestions that have proved to be popular include the non-mirror challenge; boyfriend challenges your makeup, kids' makeup challenge, and under $20 makeup challenge. Do not hesitate to challenge your own makeup, no matter how unique the challenge may be.

## Makeup dos and Don'ts

Everyone has their own horror story about their own makeup, but you can be sure not everyone posted it! Share some of your worst makeup moments with a 'dos and don'ts' clip and how you learned from them. Combining educational content with comedy is often a common combination on YouTube, and your viewers will enjoy your unique insight into makeup mishaps.

## Skincare Routine

A good skin care routine is not something that has been mastered by everyone, so if you can offer any great tips on how to improve your body, your viewers will applaud you. A skin care routine video will show each product with an accompanying summary, similar to a report of the makeup products that you use. These skincare products should be used in a live demo and expert tips on how to apply them to your YouTube viewers.

**Makeup Haul**

A good skin care routine is not something that has been mastered by everyone, so if you can offer any great tips on how to improve your body, your viewers will applaud you. A skin care practice video will show each item with an accompanying summary, similar to a report of the makeup products that you use. These skincare products should be used in a live demo and expert tips on how to apply them to your YouTube viewers.

**Game Play Streaming**

Streaming game play is one of the most popular video ideas around YouTube. New gaming channels for YouTube are popping up every day, making it a highly competitive space. Try to make your gaming videos as entertaining as possible because of the huge competition you're going to face. Give your post a humorous twist and give the unfolding action your own unique take. That's the popularity of streaming game play; it even has its own dedicated Twitch.tv platform.

## Highlights in Games

No matter how much time and money are put into games, glitches are often impossible to avoid. We can be irritating for players at times, but we may be amusing or genuinely helpful at other times. Write a video about it if you think you're the first to find a bug in a new game. Normally, game developers will address these holes with updates, but your fellow YouTube users will need to share their experiences with them while they last.

## Walkthroughs

Have you ever been trapped in a game level that you can't complete? Many people say yes, but if the response is no, then you should build walkthrough videos! Build playlists for different games and use tutorials to fill in these playlists for different parts of each game.For example, for individual levels or for different game modes, you might break your videos into walkthroughs.

## Customized Developer Options

Avatar customization is all the norm these days, with game designers acknowledging the tremendous demand for customizing hairstyles, vestments, weapons, and clothes. Step into the act by sharing your reflections in some of your favorite games on various looks.

## Game of the Year

Show viewers on YouTube how authoritatively you can talk about games in different genres by deciding on your games of the year. Many genres you might look at include first-person shooter, RPG, real-time strategy, MMO, MMORPG, racing and games of sports. If you're enthusiastic about all kinds of games, regardless of their style, making a YouTube video like this may be surprisingly easy.

## Glance at Obscure Games

Delve into the world of dark games and share with your YouTube subscribers some of your discoveries. A great game will sometimes slip under the radar and be relegated to history. But if you succeed in finding out an incredible title that is a great play, you can get lots of kudos for it. Look at games that have been released in other countries or games that have been released to get some inspiration in the shadow of more popular titles.

## Gaming News

Channels like IGN News and GameNewsOfficial, when it comes to providing players with the latest news, will undoubtedly rule the roost. But by adding some of your own personality to your gaming news videos, you can compensate for what you lack in reach and information. For gaming coverage, there's a lot to say that comes from just a daily gamer. You have no responsibility to be unbiased and can speak as much as you like about garbage.

## Talk on Old Games

Old is new as far as games are concerned. Spark nostalgic feelings by making old games gameplay videos or reviews. You could think about what made those retro games so successful and compare them with modern versions. Some video ideas on YouTube might be to share some unknown history of classic games or give some background on how it was coded.

## Answer Viewer Questions

If you can show your viewers how comfortable you are with sports, they will naturally seek feedback from you. Encourage your viewers to send questions at the end of one of your previous videos and then create a new video to read and answer those questions. If you still don't have enough subscribers to ask questions, try to ask your own questions.

## Showcase Game Modes

Changes made to fans ' games will change the dynamics of the game completely and often lead to exciting or unexpected scenarios. Although you may not be able to create your own game mods, you may be able to trawl through YouTube and convert your highlights into a highlight reel. Mods are not usually released to a wider base of games, but this does not dampen their appeal.

## Discuss Future Releases

If you're excited about games that have been announced recently, don't hold it in! To share your thoughts, make a YouTube video. We will explain how the new game will be different in the series from previous titles and look at the new features it will bring. If the game you want to talk about hasn't been released much data, you can always guess and ask what you'd like to see included.

## Making of Short Films

Upon testing your YouTube account, you will be able to upload clips that are longer than Fifteen minutes.Which means you can start uploading to YouTube your short films. To get started with some simple images, a camera, a microphone, video editing software, and a script are all you'll need. Once you're used to making short films, you can start using professional equipment and take care of yourself.

## Favorite TV Actors

Make a list of your favorite movie or TV characters and share it with your subscribers on YouTube. Count down from a number before you enter your favorite character and explain why you most like them. You can also turn these YouTube video ideas on their heads entirely and instead make a list of characters that you most hate.

## Analysis of TV Scene

Has a scene come as a surprise in your favorite TV show's latest episode? If so, why not offer a thorough analysis of what happened and explain what that means for the show as a whole. For a newly released movie or a beloved classic, you could do the same. On first exposure, people often miss things, and you can be the one that fills them in the gaps.

**Season Recap**

As a new TV show is about to be launched, viewers will try to recapture the previous season to ensure that they correctly remember it. A TV series is often going to create their own recap video, but a fan-made version can still have a lot of views. Hone in on some events that have occurred in the last season and explore their effect in the new episodes.

**Favorite Movies**

The taste of everybody in movies is different; so, the lists of favorite movies by two individuals will rarely be the same. There are many different ways to present this type of content, including' favorite 80s movies," favorite childhood movies," favorite sci-fi movies,' etc. Render visually appealing your YouTube video by showing iconic moments from every movie you're thinking about.

**Favorite TV Shows**

Another idea for YouTube videos in the same vein as favorite video movies is the favorite video shows on TV. Encourage your viewers to share their views in the comments section of the video and talk about their own favorite TV shows. Come up with different list titles and explain why you like each TV show so much to your subscribers.

## Actor Bio

In a dedicated movie, profile your favorite actor who pays tribute to their career and accomplishments. Talk about the actor's influence on you and which films you've enjoyed watching the most. Focus on some scenes that resonate with you and ask for their feedback from your audience.

## Film Trivia

To generate fascinating facts, a movie doesn't have to be amazing; The Room is a prime example. The controversial film by Tommy Wiseau, also called the worst of all time, has gained a massive following of the cult. Try to think of movies, whether good or bad, that there are plenty of stories about it and turn it into a trivia clip. On sites like IMDB, there are often plenty of great trivia you can use to help you make your video.

## Recreate Famous Scenes

Get your camera out and recreate some of the most famous scenes in the cinema. You could have only a small fraction of the original budget, but this is all part of the appeal. Design your own costumes and play different characters with your family. These videos often very quickly turn into parodies, so don't take things too seriously. In fact, a funny version of a film scene is likely to get views more likely because it offers a completely new perspective on the scene.

**Favorite Quotes**

Share a YouTube video showing your favorite quotes from the film and explain why you've selected each. The quotes that you include may be a combination of inspiring, motivational, funny, and sentimental, or even just quotes that you like for no reason. YouTube video ideas are easy to create, such as favorite quotes clips. Design Wizard has a great selection of inspiring quotes templates you can use in your video to get you started.

**Film Technology**

If movie engineering is your passion, you can make use of many different ideas on YouTube videos. You might be able to show some of the tech you have at home and how you use it to make your own movies. For example, you could look at cameras, mics, video editing software, etc. Another option is to explore how some movies or TV shows have been made. Let your subscribers know how different effects have been produced or show which lighting has been used.

**Upcoming Releases**

Tap into the hype of upcoming releases by making a YouTube video that will send you some predictions. Do some study on what the next film or television series might expect and then offer your own unique viewpoint on it? YouTube users will be excited about constantly tuning in to your channel to see if you have any new information for them.

## Sports Live Events

Covering live events can be a perfect way to attract audiences who don't attend, but want to keep up with what's going on. While you are likely to be banned from broadcasting the event's video, you may be able to film conversations with friends, half-time reflections, or experience your game day. You could make a vlog to record how you feel about it after the activity is over. There is a better chance for a passionate and genuine video to resonate with fans.

## Sports News

Connect viewers with the latest news from the sport. It might be sports news in general, or you might focus on a niche area in which you feel you have some level of expertise. While large news outlets are usually going to get most of the scoops, you can still provide your own take on the news. Keep an eye out for stories that can come directly from athletes themselves on Twitter and Instagram.

## Trick shots / Skills

Whether your favorite sport is basketball or pool, you can quickly go viral with impressive trick shots. If you think you've got what it takes, setting up a shot would be hard to achieve and record for most people. Even if you're very talented, it may take a while to get right with this YouTube video idea.

You might just be able to show off some of abilities you can practice instead of trick shots.

## Workout Advice

At some point in their lives, everyone has been searching online for fitness tips, but they may not have found the answers they are looking for. This is where the video of your workout advice comes in. Share what worked for you in the gym with your subscribers and teach them how they can put it into practice in their preparation. There are so many fitness videos on YouTube, but you can tailor your content to them if you target a specific audience.

## Diet Plan

Many diets may not work for everyone just like a workout. Nonetheless, you may have a diet plan that you find worth sharing up your sleeve. Break the diet plan into separate videos and gradually introduce it in the form of a playlist to your subscribers. If you eat those meals, instruct your viewers how to prepare them and highlight their nutritional value.

## Favorite / Best Players

Unleash your inner motivational speaker and start exercising with your subscribers in the mood. Share some of your favorite motivational quotes or write a script you think may have the desired impact. You will find it overwhelming to inspire a bunch of casual YouTube users, but if your video is presented well, communicating with your target audience will be much easier.

## Motivational Video

Get on YouTube after the big game is over and share your thoughts about what's gone down. Did your team make an amazing overall victory odd, or did they suffer an embarrassing loss? Win, draw or lose; game evaluation videos can provide fans with relevant, up-to-date content that is both entertaining and informative.

## Match Analysis

There are always hot topics for football, soccer, basketball, baseball and ice hockey, trades or drafts. There is considerable anticipation around the draft time as fans look forward to seeing what players will sign up for their squad.

## Discuss Transfer / Drafts

A good video idea for YouTube may be to create content for YouTube users who want to work out but don't want to pay for membership in a gym. Find alternative training equipment for gym weights, such as food bags, logs, water bottles, paint cans, and books. Not only do your videos offer people a new way to get in shape, but they also fascinate casual viewers who might be interested to see what your workouts entail.

## Use Alternate Workout Equipment

This YouTube video idea aims to provide the best way to stay fit for people who are unable to afford the gym or expensive diet plans. Take a look at cheaper alternatives to common dietary foods, explore home workout techniques, and encourage local park visitors. Cover your own efforts to fit outside the gym and share with your subscribers your results.

**Keeping Fit on the Budget**

Although you rarely get to put in some heavy workouts, you can still do some exercise level. You can show your audience the right way to stretch their legs under their desks while sitting down and stretch their back. This idea might be wisest to photograph at home unless you're at a lunch break! Get all the equipment you'll need.

**Before / After Workout Video**

There are few better ways to prove a successful workout than a video before and after. Showing your own story before and after success will give you a lot of credibility on the YouTube channel. Your viewers will quickly realize you know what you're talking about and can help them achieve their goals of weight loss.

**Supplements Usage**

Finding the right nutrients for a beginner can be a hassle. There are so many solutions out there that the best ones can be difficult to discover. Create a video that shows what you've used in addition to your own workout routine. Discuss each supplement's desired effect, its protein content, and how you felt it benefited you.

## Fitness Apps

Research the best available fitness apps and share your findings with your subscribers to YouTube. See what apps are great for monitoring your workouts, logging your gym sessions, tracking your calorie intake, etc. Bring some variety to your list by uncovering some new fitness apps that may not be well known.

## Fitness Myths

By debunking certain fitness theories and instead showing viewers more efficient methods, improve your credibility. This type of YouTube video can be entertaining to your viewers as well as highly educational. Tell your audience about their experience of using the approaches you spoke about and invite them to share their story.

## Song Cover Videos

Impressive coverings of songs can get very popular quickly. On YouTube, Justin Bieber, Shawn Mendes, and 5 Seconds of summer all had their break performing covers. If you can build a fanbase on YouTube, it might give you a springboard for a music career. You can do an acoustic version of a popular song or even use your own remix to put a whole new twist on one.

## Write your Own Song

If you're blessed with the ability to write your own songs, record singing them on YouTube and sharing your video. If you can also play an instrument, the better make a playlist of all your original songs to make it easier to pick them up in searches for YouTube. Post your videos on social media to let everyone know about the new song you've written.

**Music Lessons**

If you can play an instrument, providing music lessons is one of those YouTube video ideas that are easy to put into practice. Guitar lessons on YouTube are the most searched for music lessons, but many other types of lessons are also required. For these types of videos, a typical method is to split the song into parts and work step by step through them. Then you can play the whole song to show what the finished version will sound like.

**Highlight Song Similarities**

Some great songs will sound like someone else unintentionally. It can make a video in which you show these parallels for interesting content, whether it's a similar chord progression or identical lyrics.When you look at older tracks, you could find that in newer recordings, some of the beats were used again. You might also highlight a sampling of older music that may not be familiar to your audience. For example, Drake's hit song ' One Dance ' samples British artist Kyla's song' Do You Mind (Crazy Cousinz remix).

## Experiments with Unique Instruments

Expose your YouTube viewers to and play with some unusual instruments in your music from around the globe. The use of these unique instruments to cover famous songs could yield some very entertaining results. A theremin, hydraulophone, stylophone, glass harmonica, and a didgeridoo are some examples of unusual instruments.

## Checking your Musical Influences

You may find it easy to wax lyrical about the musicians who inspired you to begin playing music, so why not make a video about it. Share a story about how you were encouraged to take up the guitar, bass, or drums with a particular song or performance. In order to keep the discussion going, ask your viewers to tell you their musical influences in the comment section.

## Exploring Local Music

Not every talented musician receives the attention they deserve, but to change that, you can do your bit. Attend shows in your local area and feature musicians who wouldn't be popular with a wider audience. Interview them and upload some clips to YouTube after their gig. You could do some cross-promotion with them if they have their own YouTube channel.

## Cover a Musical Event

Many of these video ideas on YouTube are a lot of fun to do, but nothing more than covering a musical event. Travel to a festival and begin to review live performances, interview participants, and sample different stalls. You can compile the funniest moments from your attendance interviews or create a highlights video for the festival.

# Chapter 8: YouTube marketing in main sectors of life

## 8.1 Education

**What is YouTube Marketing in Education?**

There's almost no popular social media channel that's ignored as much as YouTube is when it comes to marketing your higher education institution. It's a shame because there's one of the largest and most active social communities on YouTube–teens visit YouTube 14 times more often than Twitter, according to the recent study by The Intelligence Group. Shortened attention spans increasingly require material that can be viewed easily, and short video can better fulfill this immediacy requirement than any alternative. This means you might be missing out if you don't feature impactful videos prominently on the website of your college or university.

YouTube is a powerful marketing resource that you should certainly use to the full, from sharing videos of student council elections or recreational sports to keeping current students speaking, to sharing messages from your academic faculty or administration to advertise you're various programs and attract prospective students.

**How YouTube marketing in education works?**

The first and most important step in exploiting the social media marketing power of YouTube is to create a YouTube channel for your organization where all of your videos will be uploaded. This will make discovering the videos you are sharing much easier for people as they can quickly subscribe to the channel and receive updates when uploading new content. It's also a smart idea to create different channels or site sub-sections for different aspects of your school you want to advertise, such as student life, academic programs, or opportunities to study abroad, making it easier for people to find the content that is most important to them.

Taking an example of McGill University, that holds a large variety of YouTube accounts and variable type of contents as well. Through subscribing to their McGillRezLife account, prospective students can see what life in residence is like, find out which curriculum suits their needs through checking out their McGill sub-section programs and hearing current students and professors speak about what makes them special, or see what they can expect from campus life with the Campus Life sub-section.

Dividing your content into unique categories on your channel or using multiple channels for the many initiatives of your university is a great way to ensure that students can easily find what they are looking for while at the same time giving each branch of the university you are trying to market their own identity.

## Student Participation is a key to success

When it comes to marketing the school through YouTube, using the student body's voices is an absolute necessity. Getting students to upload content and take part in your video promotions is a great way to reach out to other students because they appear to connect with people of their own age in similar situations even better. We sound less like marketing to them or selling something to them. A prospective student is much more likely to trust and be engaged by a first-year student than a member of the faculty because, on a personal level, they can relate to them.

Concordia University has a YouTube channel that accepts submissions from students studying abroad, shares some of their amazing experiences and what they have learned through the process. Not only does it offer opportunities for prospective students to study the market abroad, but it also offers students the opportunity to feel like they belong to a larger community and that they have a voice to share their stories within that community.

Alberta University has an ongoing series of YouTube videos about a single international student studying in Canada. This offers students the opportunity to speak about their college and Canadian experiences, acting as an effective tool for recruiting new students, as well as making international students feel at home while studying away from home.

## Diversifying Your Content

While showcasing your school's many exciting programs and beautiful campus is great, students are liable to get bored pretty quickly when they realize that your YouTube channel is only being used to talk up what makes your university so great. Creating smart, original, and diverse content is absolutely necessary to build a natural and realistic engagement with your students. If you just upload professors' video after video describing why the university is so amazing, students can finally see it for what it is–thinly veiled ads. You will be able to attract a much larger audience by showcasing great original content.

Concordia University regularly uploads performances on the YouTube channel of the university by their music and theater graduates. Videos such as this performance by one of the chamber music ensembles at the university are unique and engaging pieces of material that will catch the eye of many different demographics than a video of the English department chair explaining the ins and outs of the programs.

## Linking Landing Pages

One of the largest missed opportunities for higher education institutions is not to provide links on their videos to landing pages or calls for action (CTAs)–either embedded in the video or on the video page. You wouldn't believe how many colleges on their website do not include a reference to a relevant page.

While videos have great potential for lead generation education, without a proper call for action, you will never convert any of those leads.

When you have a "TrueView" PPC project running (in-stream videos as ads, paid only when the consumer watches for 30 seconds), CTA overlays, text advertisements shown in your videos can be shown in all the videos on your channel at no extra charge. For example, if you have a fantastic video explaining why your bachelor's degree in philosophy is so unique, why not provide a connection to where a prospective student can sign up for more details or contact the admissions office?

This video on the YouTube channel of Dalhousie University about the language programs they provide—part of an ongoing series produced by the Faculty of Arts and Social Sciences to promote their various programs—is included in the video description not only a reference to the program page of the Faculty of Arts and Social Sciences, but also a link embedded in the video at the end of an article on how the Russian Faculty of Arts and Social Sciences These are targeted calls for action that encourage the next move for interested parties.

**Being Active**

If you're not going on it, there's no point in having a social media account! This is especially true with YouTube, as a channel with a handful of videos that hasn't been updated in over a year is not going to do you any good, in fact it may hurt your school's reputation, as people will think that you are not active on social media, and thus either behind the times or reluctant to make a foray into what is now an integral part of students' lives. But, that's not to say you just upload material for uploading purposes. Establish clear goals to ensure your communications align with larger student marketing strategic objectives.

## Benefits of watching Educational Videos

On YouTube, we have lots of educational videos that you can view at any given time in the comfort of your home. YouTube does not charge the content they watch from their viewers. Nevertheless, it gets revenue from businesses that rely on this site to advertise.

It's part of the money they're using to play YouTube video holders. On the YouTube website, you can watch educational videos in different categories. This research explores some of the benefits of viewing educational videos.

## Values in Comprehension

One of the advantages of viewing educational videos is that learners have a better understanding of a definition. We may not know a theory well at times when learners are in class, but they get a better understanding of the subject when they get home and get to watch the videos on the subject.

Students watching educational videos are doing well in school relative to those who are not. Students are encouraged to watch videos of learning as they can learn at their own pace.

**Power of Retention**

Most people tend to forget long explanations of different issues. This is due to the so-called abundance of data. It is recommended that you download and watch videos offline as the learners can now go through the lecture at their own pace. If they need to repeat any unclear section within the clip, they will review it again and again.

**Content Mastering**

Watch educational videos if you want to master content in your study area. The simpler a definition will be to watch more and more videos often. People who repeatedly listen to music are typically sharper in the lyrics than those who don't. This also refers to educational videos, the number of times you watch the videos can help you better understand the concept of programming if you are learning how to program.

**Effective**

One of the best and effective ways to get your points across is visual contact. Learners would never forget details in visual ways that was passed on to them. This is what learners will be given when watching instructional videos offline. You can view some of these videos on YouTube when you're in school and download them using the YouTube downloader. Once you're back, during your private study, you can watch the videos.

## 8.2 HealthCare

**What is YouTube marketing in Healthcare?**

YouTube is one of the largest websites and search engines in the world, which is why it is the tool for successful marketing in healthcare. This gives healthcare brands and companie's unprecedented access to a world of current and future clients, as well as consumers and supporters, with so many people on the video viewing stream. YouTube provides a perfect platform for information, entertainment, and advertising purposes to share online videos. With all these advantages in terms of health care, it would be wise for more medical organizations to continue using YouTube to reach and communicate given customers. EnveritasGroup.com clarified why YouTube for healthcare is so amazing.

**How YouTube marketing in Healthcare works?**

YouTube is a very transparent and large network of videos that can be shared and accessed wherever possible. It is a double-edged sword for some. Most businesses, not exclusively in the medical industry, have an issue with the recommended videos from YouTube that will automatically pop up after a video is complete. This used to be a problem with a hot button, particularly if the videos were appropriate for age or not. YouTube has recently fixed this issue. Users will edit the HTML code and insert links into videos they would like to suggest before posting their video at the end of their post. Now users have the control to give suggestion of other videos that they themselves have published or to market a sister organization.

If you're in the industry of healthcare, think about expanding your reach through YouTube. It is a completely free resource that can be used by anybody, but if it is to be used by a reputable medical organization, the videos must be of high quality and professional standards.

The Mayo Clinic is a great example on the YouTube site of a medical institution that stands out. There are videos about everything related to health at the Mayo Clinic. With several playlists, the Mayo Clinic organizes its profile by topic or class. This is primarily a good tool for users to easily find a particular video. Playlists are a important part of a successful YouTube channel when an organization plans to make lots of videos on various topics. Playlists are like folders on a desktop computer: anything can be called and videos inserted on the subject in the playlist. This is especially useful if users want to share the whole playlist.

YouTube is a great way to connect with their clients for medical professionals. YouTube gives you the ability to humanize yourself and your profession as a doctor, displaying empathic, academic, and communication skills. YouTube is a great tool for training clients. This is used throughout the world in classrooms to train young and old people. Didn't you ever look for something to do on YouTube? There is an explanation for 925,000 YouTube videos showing step by step how to tie a necktie; people learn by seeing and hearing.

## Pros of YouTube Marketing in Healthcare

Video is useful for every healthcare department. It allows doctors to speak to patients directly. This helps show the distinctions of hospitals. This helps brands in healthcare to describe their services and encourage them. This helps to draw donations to medical non-profits. It allows timely information and patient education to be provided by medical professionals. And showing these videos on YouTube has a wide audience.

## No Appointment Necessary

Part of YouTube's appeal is how anyone at any time can scan, access, and share its content. This provides healthcare brands and companies with round-the-clock access to more clients, consumers, and reputation by providing people with open access to these images. And as these videos can be viewed free of charge, it helps people to discover and communicate to healthcare providers with optimum convenience and comfort.

## A body of Patient Education

The multimedia appeal of YouTube makes it an ideal way to educate patients and exchange important health care knowledge for medical professionals. For schools around the world, it is actually being used to teach people of all ages on all sorts of healthcare topics. This requires medical professionals to provide visual and verbal examples in order to facilitate, enhance, and accelerate training. Best of all, to replay and revisit the clips, people can keep coming back.

**Everything Viral is good**

When a video goes viral, it is considered as the biggest sign of success on YouTube. This occurs when a video is viewed and shared by a huge number of people randomly. While there is no way to predict which videos will go viral, by creating innovative, convincing videos that resonate with audiences, a brand or company will increase the likelihood of this phenomenon.

YouTube is a common forum that, for little effort or cost, offers outstanding visibility. This is why everyone in the healthcare industry should consider seriously increasing their scope and results by including YouTube to their marketing efforts.

**Cons of YouTube marketing in Healthcare**

**Annoying advertising**

YouTube isn't the right platform for you if you're against placing promotional content on your videos. The free cost of a downloaded video, except by changing your subscription plan, has its dark side in advertisements you can't remove. These ads can interrupt your video's message, such as this ad for online press releases. The ad for a rival with an enormous discount offer put on your video can also lead a potential customer to the website of your adversary.

**Poor customization and usability**

The website design and branding of your company will not be reflected on your YouTube page. YouTube does not endorse the integration of your logo or CTA button into a video player, and the site provides a few model models that you can use to differentiate between other organizations.

If you want to meet an exclusive B2B audience, it is best to choose another site for video hosting. Some advertisers, such as Creator Studio, may have problems finding and using usability tools. To control your screen, it may take you several minutes to reach the most prominent features.

**Low security and high wait time**

YouTube cannot prevent you from illegally downloading your video. If anyone steals your content by posting it as their own, it may affect the number of views and the SEO rankings of your website. In less than a few minutes, I found six YouTube downloader.

Lack of security isn't the content's only downfall. YouTube viewers can incur long waiting times during high-traffic bursts, so it's no surprise that after buffering for more than 10-20 seconds, they stop watching your video. It can also be a time-consuming process to upload a video. Uploading a 15-minute video can take up to an hour, adding to the packed schedules and tasks of marketers.

## Breaches of Patient Privacy

Concerns about the use of social media by HCPs also concentrate on the potential for negative impacts arising from violation of patient confidentiality. HIPAA, as amended by the Health Information Technology for Economic and Clinical Health (HITECH) Act, governs the authorized use and disclosure by covered entities, including HCPs and hospitals, of patient information. The HITECH Act outlines the notification standards for privacy breaches and expands specific provisions to include business partners. Section 13410(d) deals with civil and criminal penalties for infringements based on the nature of the violation and the resulting damages. Although social media use is not expressly mentioned, under HIPAA and HITECH these tools can certainly present risks. An HCP can infringe federal HIPAA / HITECH or state privacy laws in a number of ways when posting to a social networking site information, posts, photos, or videos about a person. When interacting with or about patients on social media, infringements of patient confidentiality will lead to legal action against an HCP and potentially its employer.4 Moreover, it is important to note that

HIPAA does not prohibit the dissemination of "de-identified" medical information.

In 2003, the Department of Health and Human Services (HHS) released the HIPAA Privacy Act, which lays out the first national privacy requirements to be enforced for "covered entities," such as HCPs, hospitals, and health plans. The HIPAA Privacy Rule imposes heavy fines and potential criminal charges on the unauthorized disclosure of individually identifiable individuals.

In order to comply with the HIPAA Privacy Act, clinical vignettes shared on patient-related social networks must have all personal identifying information and any compromising links removed. This "de-identification" can be achieved by altering or omitting key patient details (e.g., names, insurance or social security numbers, date of birth, and photographs) and preventing the disclosure of rare medical information.

Consent from the patient is a critical issue that needs to be considered when using social media. Through consideration of the place of publication, an HCP or health care organization can decide if patient permission is required. Using unique HIPAA-compliant messaging systems, such as those given inside Doximity, can be potentially secure even for patient identity information, provided the recipient has a medical justification for accessing such information. However, before posting de-identified case details online, it is eventually up to the individual HCP, the exercise, or the organization to determine when they will seek patient consent.

## Violation of the Patient–HCP Boundary

While patients initiate online communication, HCPs that communicate on social media with their patients may breach the patient-HCP boundary. A recent study found that on Facebook, patients often extend requests for "friend" online to their doctors. Nonetheless, very few doctors reciprocate or respond, as it is generally thought that an HCP is ill-advised to communicate with a patient through a general forum on social media such as Facebook. Furthermore, statements of organizational policy often prevent private online communication between HCPs and patients.

Therefore, HCPs should become familiar with the privacy settings and terms of agreements for the social media platforms they belong to so that they can maintain strict privacy settings on their personal accounts.10 Instead of being "friendly" or interacting with a patient on social media, HCPs may recommend that the patient should set up a website specifically designed for medical messages.

Physicians may also breach the personal boundaries of a patient by improper use of information found online or on social media. Since social media can provide a great deal of information about patients, it can be used to support clinical care in a positive way. This method, known as "patient-targeted Googling," has been identified in many medical settings. Anecdotal reports have highlighted this.

An HCP may detect images or photos on social media sites showing patients engaged in risk-taking or health-averse behaviors. Online examination of patients ' personal activities, such as whether they have quit smoking or maintained a healthy diet, may compromise the confidence required for a good patient-physician relationship. In such situations, therefore, an HCP should be considered.

**Licensing Issues**

The use of social media may also adversely affect the qualifications and licensing of an HCP. State medical boards have the power to punish physicians, including the enforcement of limits or the suspension or revocation of licenses. Such sanctions may be enforced for unprofessional behavior, such as improper use of social media, sexual misconduct, infringements of patient privacy, manipulation of prescription data.

U.S. licensing officials have reported several ethical breaches by HCPs on social media, which resulted in disciplinary action, for instance, a Rhode Island State Board of Emergency Medicine physician was reprimanded for "unprofessional conduct" and fined after posting on a patient on Facebook. The doctor did not mention the name of the patient in the post; nevertheless, it was enough.

Nursing boards have also punished nurses for violations involving the electronic disclosure of personal health data to patients and have imposed sanctions ranging from letters of concern to license suspensions. It is also fairly common for HCP students to post unprofessional material on social media. One survey found that 60 percent of deans of medical school reported incidents involving students on social media.

**Legal Issues**

Social media's pervasive use has brought new legal complications. The use of social media can be applied to a number of constitutional rights, such as freedom of expression, freedom of searches and seizures, and the right to privacy; however, these rights can be successfully challenged. A U.S. in 2009. By making derogatory remarks about the color, gender, and religion of patients under their care, District Court upheld the expulsion of a nursing student for breaching the honor code of the class. The court concluded that the school's honor code and privacy agreement were signed by appropriate standards of behavior controlled by each nursing student, denying the student's argument that their right to freedom of movement was violated.

Legal cases should never be shared on social media because most current case law dictates that such data is "discoverable," although this may rely on the reason for which the information is sought. Although posted anonymously, numerous forensic techniques may be used to tie legal information directly to a particular person or event.

HCPs may also be exposed to litigation if they respond by providing medical advice to a question sent via social media. In response to requests for such guidance, it was proposed that a legally sound solution would be to provide a standard response form: advise the inquirer that the HCP does not answer online questions; provide offline contact information so that an appointment can be made if desired; and provide a source for emergency services if the inquirer cannot wait for an appointment.

**Damage to Professional Image**

A major risk associated with the use of social media is the sharing of non-professional content that may unfavorably reflect on HCPs, students, and related institutions. Social media communicate information about the nature, beliefs, and interests of a person, and the first impression created by this material may be lasting. Perceptions may be based on any of the data in social media.

Behavior that could be viewed as unprofessional includes breaches of patient privacy; the use of profanity or discriminatory language; photos of sexual suggestiveness or intoxication; and negative comments regarding patients, an employer, or a school. These public errors have been reported by HCPs, including physicians taking electronic photographs throughout surgery, posing with arms.

Information collected from social media can also be used to make decisions about admission to medical or educational programs, residential placement, or employment. Employers and residency programs are now looking for Facebook and other social networking sites before recruiting applicants. A survey conducted by Microsoft found that 79 percent of employers view online information about prospective employees, and only 7 percent of job applicants were aware of this possibility. By making public posts, a person has willingly made information available to anyone for any purpose. For some, it follows logically that candidate who does not use discretion to decide what content should be posted.

Ideally, account and privacy settings should be configured in a way that allows one's network to broaden while restricting data exposure to people outside the network. Any settings made available by the social media site that allows users to identify different relationships so that only appropriate information is shared with certain groups or individuals should also be used.

## 8.3 Agriculture

### What is YouTube marketing in Agriculture?

Recent studies have shown that farmer's use social media use it almost every day. YouTube is one of the most popular social media sites, and it said more than fifty percent of farmers use this YouTube to learn about general news, enjoy music, or Agricultural Videos ' how-to-video.

Also, in the agricultural industry, social media is prevalent. There are forums, Twitter pages, Facebook groups showing what's going on in agriculture. Such social media platforms will share the real story behind agriculture.

Such social media platforms allow those who want to learn a lot to communicate with other people in the industry and learn more about costs, land management, new techniques, and methods.

Yesterday, Bloomberg cited a 2018 study by the Pew Research Center that found YouTube to be among those in rural America's most popular online platform— at 59 percent— with more eyes than Facebook, Instagram, Twitter, and more. Social media manager Keith Good, who works for the University of Illinois farmdoc campaign, said to Bloomberg, "Farm associations and commodity companies have urged producers to join the social media conversation."

Social media is commonly used and is very useful for collecting relevant information about the farming industry. Social media is a major factor in improving productivity because new agricultural learning concepts become accessible, and it helps you to communicate with others who do the same. Farmers use social media to reach out directly to ordinary people.

Technology has changed the methods of farming as well as the means of communication through which people learn about farming. A new audience was created by social media.

Farmers and agribusinesses are able to tell their stories and express what matters to them. If there is internet access, social media is accessible at the disposal. The group will engage in discussions and collect ideas and perspectives.

## How YouTube marketing in Agriculture works?

Agriculture and social media have a lot in common — cultivation is the biggest commonality. Before plunging into the sphere of social media, here are some things to consider:

## Choose your Crop

You need to choose the best network that fits your needs with so many different social media channels out there. Every website serves various audiences and provides various types of posts. You may have short news stories that can be tweeted on Twitter or an impactful video better posted on YouTube. Finding out who you want to connect and what content you want to share will help you figure out which social media platforms are right for you.

## Cultivate your crop

You can't just plant and expect your seed to expand. Social media is the same case. You will keep it filled with meaningful content when you create a Facebook page or Twitter account. Each photo, article, blog post, or video your company will make in the future will continue to generate page views and awareness. Share farm messages, post your process clips.

## Give it Time to grow

Following the development of a strong social media takes time. Next, you need to "listen." Observe online conversations through social media networks in order to maintain a consistent and up-to-date understanding of what is important and of community interest. Then you can incorporate material that resonates with your target audience based on the "listening" you've done.

## Pros of YouTube Marketing in Agriculture

In an era when a billion people start conversations on Facebook, 320 million people check the latest news on Twitter, and every minute 300 hours of new video is uploaded to YouTube, social media gives a voice and influence to agriculture.

The rapid development of social media platforms enables the specialty crop industry to talk directly to the public, educate customers about food production, and inspire them to become advocates of agriculture. Whether you're using Facebook, Instagram, or Twitter, you're now able to publish your own stories without going through previous gatekeepers— the press.

Recently, Pew Research revealed that over 30 percent of Americans are using Facebook as their main news source. That's 96 million people who could hear your post theoretically. As social media sites become the news powerhouse, the question should no longer be: "Is social media right for your business?" You should rather ask:' how can I use social media for my buses in the right way?

**ADVANCING AGVOCACY**

As Chipotle's E and the Golden State dries up. Coli outbreak alerts consumers; social media provides a channel for farming to educate the public on hot topics such as water supply and food safety. The dialogue about food, science, and agriculture can be influenced by online interactions. Most notably, it is possible to combat misinformation on issues surrounding farming and nutrition.

Around 98% of the U.S. population is not producing food. Such consumers need first-hand experts to help them understand the farming practices of today, where their products come from, and the regulatory concerns that could potentially affect the food they eat.

Sharing stories and important food production facts will inspire customers, food companies, and other farmers to get interested in farming issues. A story can go viral in seconds via social media, and the effect is tenfold. You can share your side of the story with one tweet on Twitter and potentially create a whole host of supporters for agriculture, or advocates. Social networks allow you to connect with the public — before someone else.

Activists are gradually starting to harness social media's power and use its effect to influence their audiences. For example, Erin Ehnle, who grew up on the corn and soybean farm of her parents, launched a Facebook page called "Keep it Real: Through the Lens of a Farm Girl," where she posts photos of her own farm, built with hard-hitting farming information. Her profile had the first week's 200 likes, and the 10-day mark was 1,000. She now boasts nearly 30,000 site loving Twitter.

"There's so much disconnection between consumers and farmers and so much negative about farming, so my aim was to get the attention of consumers," Ehnle said. "I've shared some divisive pictures of GMOs, sustainability, and how far we've come, and some people are going to go off on pesticides rants or anything personal. But the farming lifestyle and hard-working farmers seem to be respected by everyone. We're the world's largest industry, and I believe we need to invest in our history and secure our future. Activists are gradually starting to harness social media's power and use its effect to influence their audiences. For example, Erin Ehnle, who grew up on the corn and soybean farm of her parents, launched a Facebook page called "Keep it Real: Through the Lens of a Farm Girl," where she posts photos of her own farm, built with hard-hitting farming information. Her profile had the first week's 200 likes, and the 10-day mark was 1,000. She now boasts nearly 30,000 site loving Twitter.

**Mastering Marketing**

Who doesn't love pictures of beautiful, colorful crops that grow under a golden sunset in never-ending fields? Platforms like Instagram and Facebook give you the ability to show customers exactly where and how they grow their food.

Research published on eMarketer shows that Facebook posted images earns an engagement rate of 87 percent from fans. No other type of post— including links, videos, and status updates — received an engagement rate of more than 4 percent. Social media can help businesses build a customer base relationship by enabling them to share their experiences through engaging photos that tell the story of their company.

Farmers' selfies— better known as "felfies"— have become mad about the internet in 2014. Consumers like the people behind the products they buy have a sense of it. The quick action of using your smartphone to snap a photo of yourself and upload it to Facebook and Instagram allows you a quick and easy way to communicate with shoppers. Therefore, social media is an inexpensive way to market your brand to consumers you have never been able to reach before. When you sell to customers directly, it's important to build awareness about what you're selling.

In 2010, the Tanimura & Antle (T&A) Facebook page, a member of Western Growers, posted a total of two likes per post. The company site today garners every post from 100 to 600 "likes," an average of 30 people sharing T&A's post on their own personal page and dozens of comments per post. (Post likes when a user gives a thumbs up to your post.) On a monthly basis, T&A Facebook posts receive twice the amount of contact with its 14,000 followers relative to companies with more than 100,000 followers.

"People like to see where their food comes from inside. They like to communicate with those responsible for food production, "said Ashley Pipkin, sales and marketing director at T&A. "If we don't show them what we're doing, they're going to make their own conclusions and opinions based on what they're reading on the internet. We like to show them everything from seed to shelf, and they seem to be the most common of these posts.

### Cons of YouTube marketing in Agriculture?

A more recent Farm Field Schools (FFS) analysis–a participatory approach to farmers ' education and empowerment with discovery learning, experimentation, and group action–found the cost per participant ranging from US$ 20 to US$ 40. However, these estimates do not take into account the cost to beneficiaries that can make the approach relatively expensive compared to other types of program approaches. While government spending on extension services has typically been low, evidence from several countries shows that extension of agriculture is a pro-poor public investment with high returns on poverty reduction. Poverty in Ethiopia has been reduced by 9.8 million from just one farm extension visit, and improvements in extension visits in Uganda have reduced stunting and malnutrition in children.

Cultural, social, and structural factors hinder women's involvement in growth. While it is still common for agricultural extension workers to be male and only communicate with men, in recent years, there has been increasing awareness of the role of women in managing farm households and the importance of female interaction in the adoption of new farming methods and technologies.

Many farmers live in a world of data that is unreliable and are subject to great uncertainty about weather conditions, insect attacks, and business choices. Increased access to information and communication technologies (ICT) could alleviate some of these uncertainties. Good, accurate, and timely information will help farmers respond better to price signals and help expand the scope and effect of extension and advisory services. For example, Safaricom Ltd and the Kenya Agricultural Commodity Exchange collect and distribute current commodity price information through a low-cost SMS provider to Kenyan farmers.

# Chapter 9: Role of Social Media on YouTube Marketing

Many people don't seem to think of YouTube as a social network, but they should. YouTube is known to most of us. To get our daily dose of news from channels like MTV or The Young Turks, recipes and cooking shows from The Food Network and SORTED or even beauty and fashion tips from the likes of Michelle Phan, LKISStyle and more, more than 84% of us tune-in at least once a month. Canadians still like Facebook but what I found most interesting about the story focused on social media was that while it included Facebook, Twitter, Instagram, Pinterest, LinkedIn, SnapChat, WhatsApp, and Tumblr, it wasn't YouTube.

I've been curious. Is YouTube a network that is social? How could this be important? If so, marketing teams will begin to include the site more regularly in their social media strategies, allowing them to capitalize on exponential growth and captive, socially engaged, and hard to find readily accessible audiences on YouTube.

In fact, because of relatively new and unique advertising products from YouTube, such as TrueView, marketers only pay when the viewer wants to view the whole commercial or the first 30 seconds of the ad, whichever comes first, and because of that they also receive a large amount of added value around each advertisement. For example, in addition to providing paid impressions, advertisers get free partial and organic views (both driving awareness and recall) in addition to commitments such as subscribers to their channels, also known as fans and shares.

In reality, top videos often get much more organic views than paying views. When potential customers are not your thing to miss the ads, there are other options, including Google Preferred, which was only introduced last month in Canada, where advertisers can purchase specific audiences and platforms, and Mastheads, where advertisers can take over a limited daily inventory such as the YouTube Homepage. Visitors and logged-in users are subscribing to sites that attract them, much like those on other social networks. They like videos that they find interesting, or they speak in YouTube, thumbs up (or down!). They will post and comment as well.

Content recommendations are based on user behaviour and social graph/connections, similar to how friends and their content are recommended by social networks. The "newsfeed" of YouTube, which shows content from the networks to which a person has subscribed, with suggestions, is very close to the "Wall" of a social network. Visitors can create profiles, upload a headshot, and follow their friends, or subscribe to YouTube-speak.

Ninety-five percent of all entertainment content views come from user uploads rather than official videos. This means that when someone watches a clip from films like The Dark Knight or Anchorman, or TV shows like Black Sails, they watch a video posted by a fan and not the official Paramount, Warner Bros, or Starz site. But don't worry; the copyright holder is still being paid because of the patented monitoring technology of YouTube.

Ninety percent of all brand-related content views are based on consumer uploads rather than official brand images. This means that 95 percent of the time someone watches a clip about a test drive from Kia or a demonstration from L'Oreal Mascara, they watch a video uploaded by a customer or expert, not the official product site.

More than half of all users are affected by YouTube purchasing intent and decisions. For example, 53% of U.S. customers say YouTube has affected their purchasing behaviour. Compare that with traditional social media, and you see some similarities, where 55% of respondents engaged on Facebook with brands, followed by 21% on Twitter, and 10% on Pinterest. While watching, one in three visitors is posting a YouTube video. YouTube can be categorized as a social network for many reasons. You may argue, in fact, that it is one of the largest online social networks. Some, like ZEFR's co-founder and chief executive Richard Raddon, even feel it's the future, having said recently: "If Facebook, Twitter, and Pinterest are part of Social 2.0, then video networks like

YouTube will be appropriately coined Social 3.0–in other words, the future."

## 9.1 Social Media Marketing is good or bad

The easiest way to connect with your fans feedback is to boost interaction. It makes them feel valued and respected when you respond to people and involve them in conversation. You are more likely to share your video and continue following your channel when people feel linked to you.

It can be extremely time-consuming to respond to each for videos or social media posts that have thousands of comments. Nonetheless, making an effort to respond or comment back to even one-fourth of them shows viewers that you're human, and you're trying to connect.

You can also use reviews from fans to receive feedback on your photos. Ask them for suggestions or improvement areas, and share it if they like your video. We are working with clients to expand our base of supporters, so don't be afraid to ask them to share their thoughts. See the Starbucks Twitter feed for a great example of audience interaction and engagement.

Not all feedback is good, sadly. Often people share opinions that we may not like to see-but to dismiss feedback is a bad idea for a company. Listen to these thoughts, rather than ignoring them. Is there something else you can do? While it's tempting to say,' well, it's just one bad comment'-the truth is that somebody who took the time to share it is a bad comment. Most consumers may simply find another company with which to do their business. Rather, see what you can know, apologize to the client, and do what you can to address their concerns.

In one of their video threads comments on Facebook, you can see JetBlue receiving somewhat negative feedback. They apologized and validated her concern instead of ignoring the fan.

Sharing Means Caring In the digital marketing environment, it drives traffic and gives you credibility to communicate with influencers and share your content with others in your industry. For YouTube advertising, it works the same way.

The promotion of videos by cross is a win-win in initiative in which both parties' profit from each other.

Check inside your niche for businesses or YouTube influencers or one that aligns well, preferably with a larger follow-up than you do. Joining a video or sharing the videos of each other on social media can drive a whole new audience to your social network on YouTube.

Don't forget about Thumbnails. A thumbnail video is like a book cover; it can motivate you to open it or move to the next option. A thumbnail is a static example of the content of your video. It's what's shared, shown in searches, channels of social media, and websites. Clean and attractive thumbnails will increase your dedication by 154 percent, according to marketer Neil Patel.

## 9.2 How to Use YouTube and Instagram to Establish Authority

Amanda is a video marketing strategist who helps people in their space to become thought leaders. She has a history in the making of films and is called The DIY Video Roadmap. Amanda offers tips to help you make videos that people want to watch, discusses how your experience can be best demonstrated, and more.

He studied film with Tim Ferriss Amanda and lived in L.A. Upon. A series of fortuitous events soon connected her to Tim Ferriss, The 4-Hour Workweek's influential author. For his book, Tribe of Mentors, Tim hired her to create promotional videos.

The freewheeling approach to video by Tim at the moment was different from the more formal methods that Amanda had studied in film school. Luckily, as a student, Amanda had worked with a number of companies and was able to embrace the more relaxed style of Tim for the diverse array of promotional videos they produced in their batched shoots.

In particular, the relative newness of online video sites like YouTube, rapid speed of technological advancement, and expanded availability of higher-level video resources left

Amanda and Tim free to explore new and different ways of telling video stories.

While YouTube remains the main strategic hub for Amanda, she also wanted to drive traffic to her YouTube and Twitter channels from Facebook, Instagram and LinkedIn. She wrote on those other sites about her YouTube videos, suggesting that no matter if people were actually watching her video content; they would still note her strong performance and knowledge topics.

Users don't always think from their personal profiles that they can connect to YouTube on Facebook, but that's not mobilizing what could be a huge established audience. At Social Media Examiner, we've found that on Facebook connections to YouTube now fill what's called the "Open Graph," offering a broad preview picture rather than a flat link, which is another great opportunity for your videos to gain attention.

YouTube is still a relatively new world, and it is used as their main social media platform by very few individuals. It's important to promote your videos on Facebook and Instagram because if you're just posting on YouTube, you're losing a huge part of your audience online. Often, the YouTube algorithm means that even subscribers may not see every video you upload, so sharing it wherever you can be vital.

"As you authentically communicate, people begin to connect with who you are. They really want to work with you when they connect with who you are, "explains Amanda.

Amanda also highlights the benefits of Stories Highlights that can display your services, testimonials, and frequently asked questions. The immediacy and privacy of the following direct conversations can often result in a shorter journey to close a sale.

While many YouTubers repost their videos to the longer-form IGTV tab, Amanda has opted on her own channel against this strategy, preferring to drive traffic to YouTube for her videos that are rich in content.

# Chapter 10: Future of YouTube Marketing

## 10.1 What's next?

The first video of YouTube may have been uploaded on April 23, 2005, but on February 14, the domain name of the website was registered. Hence this past weekend's 10th birthday anniversary at a time when online video service from Google is more popular than ever. There's a path from zero to a billion subscribers in a decade, but what's ahead as YouTube heads into its teen years? Here are some predictions.

YouTube remains a video service in 2015, though one that is introducing an offshoot of streaming music: YouTube Music Key. Expect it to grow into a full entertainment system over the next decade, where you're not just going to watch and listen: you're going to play. Players and activities that are interactive, in both instances, lead to YouTube into the world's largest library of content on virtual reality. This is due to the fact that games are going to sit nicely in the entertainment mix of YouTube alongside shows and music. Users are now watching hundreds of millions of videos of games on YouTube and playing games in similar numbers there by 2025: a major casual gaming site with a legacy that includes Twitter and Apple and Google's app stores.

Each time the English Premier League sell off their TV rights, there are reports that YouTube will join in the bidding with the likes of Sky and BT. These rumours have not led anywhere every time—so far. However, live sports will be a key part of the video offering on YouTube by 2025. Not only football, but American football, basketball, and cricket, as well as any other sport you can name, from mainstream to niche. And not just because YouTube has conventional rights broadcasters out bidden, even though that's the most likely short-term explanation. As the barriers between traditional broadcasting and online video topple, YouTube will emerge as a partner for sports leagues to retain their rights and go direct to fans. Its relationships with US leagues, the NFL and NBA for highlights are a tiny sliver of what's possible. By 2025, every match in every league could be available on-demand, globally, on YouTube through those leagues' own channels – at least for those that have done deals with the company.

YouTube will remain a mammoth video library to watch, but by 2025, knowing your interests will be much, much better, based on everything that your parent company knows about you as well as your YouTube past.

Yes, you'll be able to look for something to watch, but YouTube's default mode will be a lean-back channel of content customized on the fly for people watching right now: it'll know that, obviously, through a combination of cameras and microphones on the watching device, and data from the nearest Nest smart-home hub. Sign up for the Media Briefing: news for newsmakers Read more A few episodes of the latest

Scandinavian crime drama; a few music videos of new artists YouTube knows you're going to like; a half-hour sketch shows automatically cut from your favourite YouTubers; a personalized news bulletin; and micro-targeted ads all the while.

In 2025, the idea of a TV rating system based on a limited sample of viewers will seem comically archaic: anyone who creates content will have real-time data on how many people have watched, but also on what they think. In 2015, several smart TVs were warning us that any private conversation can be captured and uploaded to a server. This will be the key feedback loop for digital entertainment by 2025: massive servers humming away to process our pronounced opinions on programs, advertisements, characters, and writers. Ultimately, yelling at the **tally** will have the potential to influence the people who make whatever you watch on it. While your life, politics, culture, and whatever your cat dragged last night will be just more grist to the recommendations mill of YouTube as well.

Today's Google's Hangouts app is the model for social television co-viewing's future, where everyone is a star in their own continuously running Goggle box series. We're going to watch shows, sports, and other live events online with our friends and family: an audience of silent video thumbnails laughing and crying with us and then talking about them. How will producers and brands guarantee water cooler moments of this kind? Producers and brands will guarantee moments by paying for them. There will be a rate card to bring people together to watch, from music video premieres to new dramas or party-political broadcasts. There are plenty of zeroes on it.

The biggest star of YouTube, PewDiePie, created 4.1bn views from his channel in 2014. Where is he going to be in 2025? Running one of the biggest entertainment networks in the world is quite likely if he fancies the task. The fresh-faced YouTubers of today will be the outstanding broadcasting moguls of tomorrow. PewDiePie–Felix Kjellberg–has already suggested that he believes he can do a better job to run a multi-channel network.

In 2015, Facebook will shape YouTube's domination in online video as the most realistic-looking threat. Facebook users are already watching 3bn videos on the social network every day, striking their own deals for professional content, while reportedly courting the stars of YouTube to start uploading videos directly to their service.

Fast forward, however, ten years, and this isn't an online video market: it's just video (and music and games, and virtual reality...). A market where YouTube's rivals are Facebook, Apple, Google, and Amazon, but also Netflix, HBO, and perhaps even Snapchat, if its youthful audience grows up keeping their habit. Today's companies, including Vessel, also have their own dreams to become ESPN, MTV, CNN, and Discovery's digital age counterpart–the new brands that originated from the cable television period. The dilemma for traditional television networks is how they fit in with this duelling technology company world trying to dislodge the crown of YouTube.

**Concerns or Celebrations of Future**

Not all of these predictions were made in a celebratory way. Some of these innovations might be thrilling, but some might be frightening – or at least troubling for different reasons. Would we like to turn over our entertainment choices to algorithms that are completely unknown to us? What leads with serendipity: by chance, finding things that we didn't think we'd enjoy–or, perhaps, things that challenge our beliefs? Is the privacy-personalization market worth the risks? Do we have faith that YouTube and Google will support the producers of the programs, music, and games we enjoy? And perhaps most pressingly, will all these issues be made moot by the advent of something else–perhaps dreamed up by someone who is now a teenager–that will make YouTube a historical footnote by 2025 rather than our entertainment master?

## 10.2 Video content marketing revolution

In just a few years, how did video become such a force? YouTube, which is part of the Google Display Network, is the short answer. YouTube has stretched the possibilities and limits of the video in a single decade, creating an entirely new role for the medium in the culture of today. With four billion views a day (and 300 hours of video posted every minute), YouTube has transformed the Web into modern age television–and it's a television that gives advertisers unparalleled opportunities. After their 2015 collaboration with Google Brain, YouTube marketers have been able to integrate AI and big data to help consumers and advertisers hit their target audience with precision. The numbers tell the story: The most famous 2017 advertisements on YouTube received over 25 million views, with the top spots going to Clash Royale advertising: The Last Second (110.7 million views) and Samsung's India Service: "We'll take care of you wherever you're" ad campaign (150.3 million views).

Compared to other important factors–such as easy access to affordable high-speed broadband and improved IT infrastructure, which mitigates the constant buffering and sluggish download speeds of the past–video has become a virtual juggernaut in the marketing industry, making it an essential resource for customers to search and shop. About 50% of Internet users say they are looking for videos related to a brand before purchasing it in a store in a recent study. Similarly, 90% of consumers said watching a product video would help them make purchasing decisions. Furthermore, research shows that four times as many customers prefer to watch a brand clip instead of reading it.

Although written content can illustrate an item, videos can explain it visually—and just as visually (and audibly) display the reactions of people using it. As a result, this audio / visual aspect is more realistic and personal in engaging the senses of the user.

## Marketing on YouTube

Studies show that 60% of marketers are using video in social media marketing, while 73% of marketers are planning to increase their use of video in social media. 48% of marketers are preparing to include YouTube in future marketing strategies.

Perhaps the most significant statistic shows that after watching branded video advertising on social media, 64 percent of consumers make a purchase. With its vast audience (more than 500 million video viewers a day on Facebook alone), social media provides the perfect forum for advertisers to share insightful videos that can be watched by hundreds of millions instantly and at any time.

Another advantage of YouTube is that it allows users to share the content they want—which in turn provides marketers with massive exposure to the brand at no extra cost.

## Engagement of consumer

Studies have shown that more than 55 percent of consumers are paying close attention to the videos they watch, compared to 39-43 percent averages that skim through blogs, social media posts, and news articles. Furthermore, 45% of users say they would like to see more video ads in the future.

Video has a similar impact on consumers of B2B. Research shows that, if the choice is made, 59 percent of business leaders prefer to watch video over text. Likewise, more than 80% of senior business executives report they stream more video online than they did last year. Considering these figures, it is not shocking that over the next year, 96 percent of senior executives say they plan to use video in their content marketing.

**Does YouTube marketing really work?**

YouTube is the second largest search engine that processes more than three billion searches a month. Think about what this means— after direct searches by Google, people turn to YouTube to find solutions to their problems— seeking video tutorials and other information to address their pain points. Furthermore, many vital audiences are more likely to watch it than cable TV. These facts alone show the enormous possibility that your target audience will see your video content.

YouTube is expected to grow further in the next few years, with further improvements that will favour companies that integrate it into their strategies for social media and video advertising. YouTube is expected to make some major upgrades this year, from added profile and socialization features to better connect audiences to exploring long-form, TV-style content.

There are tons of examples of successful, high-profile marketing campaigns on YouTube today. From the insightful Whiteboard Fridays by Moz and Rand Fishkin to the celebrated series of emotionally powerful animated shorts by Chipotle, there are currently plenty of brands showing how much opportunity there is to create value and recognition by promoting great video content on YouTube.

# 11: Conclusion

YouTube content needs to be developed with your viewers and the algorithm in mind in order to grow a YouTube channel. For viewers, make sure they get their attention from your thumbnails and deliver what you promised. Start with a strong hook on your videos and edit the material to keep it interesting. All in all, you want to create an experience that keeps watching the viewer.

When you invest in a marketing strategy for YouTube, remember to concentrate on your video content before even considering marketing it. With 300 + hours of video uploaded to YouTube every minute, the playground is large and crowded, and your videos need to talk to your audience to be successful. Once you've created a video, you're proud to promote it and distribute it through all of your channels and consider investing for your best videos in paid promotion.

Video is one of the world's most popular forms of content, and the truth is that it is unlikely to go anywhere soon. And it makes sense: we crave connection and personality in an impersonal digital world. In a real-life context, we want to see and hear people— it is meaningful. Not only is video enjoyable, it's one of the best ways to get close to your audience and give them a real snapshot of what you and your company or customers are doing. The key here is to think beyond profit and product— to show them something about your philosophy, or to share information about an interesting event, or to provide some valuable information. The more they know about their positive practices, the more likely they will remain.

Thankfully, it also improves the exposure of your videos and, therefore, your channel via the YouTube algorithm by gaining clicks and holding interested viewers. Interesting content that people click on can boost the visibility of your videos across YouTube, leading to a good watch time and session duration? Your chances are also strengthened through playlists, alliances, shows, and end screens.

YouTube is still considered by experienced social media marketers to be a viable advertising option. However, marketers need to keep a close eye on what's going on with the video advertising offers from Facebook. While YouTube has Google behind it, Facebook is a giant of its own and is known to take over the social marketing landscape quickly and completely.

YouTube Analytics is a treasure trove of data that can be used to enhance your videos and broaden your stream. At first, it may look intimidating, but it only takes a quick summary to help you master the data analysis of YouTube and take full advantage of all the amazing features and knowledge. Remember, high watch time and audience retention are heavily rewarded by the YouTube algorithm. So, the best advice we can offer is to keep your spectators engaged and come back for more.

This is a constant process that should never stop, even if your goals are being smashed right and left. YouTube's top performers always dig into their YouTube Analytics metrics to see what works, what doesn't work, and how to bridge the gaps.

Besides vanity metrics such as impressions, views, likes, comments, and shares, is a well-executed YouTube marketing campaign going to have a positive impact on the bottom line of your business? Mostly, how effective you can find YouTube advertising depends on how you define the effectiveness of a campaign. Like any other, video content should be produced with a firm idea of the experience of the consumer and the funnel of your internal sales. It will probably take time to build a following — as with any social channel— but there are ways to get your YouTube popularity to start a leap, such as setting up an AdWords campaign. Eventually, you will see success if you remain committed and continue to produce quality content.

Of course, the content marketing plan still has plenty of space for other news. There is always a well-rounded, active advertising strategy— but there is no doubt that YouTube ads should be at the forefront.

## 12. References

**1. What is YouTube marketing retrieved from-**

https://www.quora.com/What-is-youtube-marketing

**2. Types of ads retrieved from-**

https://blog.hootsuite.com/youtube-advertising/

https://www.marketing91.com/7-types-of-youtube-ads/

**3. Information about YouTube channel retrieved from-**

https://www.lifewire.com/channel-youtube-1616635

https://money.howstuffworks.com/youtube4.htm

**4. SEO data retrieved from-**

https://backlinko.com/how-to-rank-youtube-videos

https://blog.hubspot.com/marketing/youtube-seo

**5. Role of Social Media in YouTube Marketing retrieved from-**

https://www.clickz.com/is-youtube-a-social-network/25701/

**6. Future of YouTube Marketing retrieved from-**

https://www.theguardian.com/technology/2015/feb/16/youtube-in-2025-future-predictions

https://daily.jstor.org/how-youtube-is-shaping-the-future-of-work/

www.ingramcontent.com/pod-product-compliance
Lightning Source LLC
Chambersburg PA
CBHW070639220526

45466CB00001B/233